Textile Science and Clothing Technology

Series editor

Subramanian Senthilkannan Muthu, SGS Hong Kong Limited, Hong Kong, Hong Kong

More information about this series at http://www.springer.com/series/13111

Subramanian Senthilkannan Muthu
Editor

Textiles and Clothing Sustainability

Sustainable Technologies

Editor
Subramanian Senthilkannan Muthu
SGS Hong Kong Limited
Hong Kong
Hong Kong

ISSN 2197-9863 ISSN 2197-9871 (electronic)
Textile Science and Clothing Technology
ISBN 978-981-10-9624-2 ISBN 978-981-10-2474-0 (eBook)
DOI 10.1007/978-981-10-2474-0

Printed on acid-free paper

This Springer imprint is published by Springer Nature
The registered company is Springer Nature Singapore Pte Ltd.
The registered company address is: 152 Beach Road, #22-06/08 Gateway East, Singapore 189721, Singapore

Contents

Sustainable Textile Technologies

S. Palamutcu

Abstract Conventional textile production technologies have accepted almost the same in principle since the ancient time. Basic techniques of yarn spinning; surface manufacturing techniques of weaving, loop-based knitting, braiding, and felting; dyeing, printing, and sewing techniques; and equipment are not changed in principle. Principles are quite the same; however, it should be stated that technology improvements, especially production speed increase, are uninterruptedly continued to be able to respond the urge of consumption. Depending on increasing consumer demands, environmental concerns are also erected eventually. In textile production chain, every processing steps have its own environmentally harmful influence on the nature. Every single fiber, every yarn bobbin, each square meter of fabric, each chemical, consumer cycle time of each textile item, and recycle or waste stage of every single T-shirt leave its own footprint behind. Conventionally used natural fiber types of cotton, wool, silk, and linen have their own environmental footprint relevant to their growing–processing steps and consumer using stages. Synthetic-based traditional man-made fiber types have their footprint of raw material and degradation time in nature. New generation of biodegradable man-made fiber production technologies offer promising possibilities from the view of sustainability. Traditional yarn production technologies and processing machinery lines have not been changed in principle since industrial revolution. However, production speed of the machinery and production efficiency has been improving constantly with the cooperation of material science and information technologies. Besides high-speed production on traditional yarn production, new yarn spinning technologies of less machinery requirement and high production speed give promises to improve sustainability approach in yarn production technologies. Weaving, knitting, and nonwoven technologies are the basic textile surface production methods. Weaving process has the biggest environmental footprint, knitting process has the second place, and the environmental footprint of nonwoven production process is the smallest. Influence of wet-processing stages depends on the amount, temperature, and chemical load of the wastewater that is discharged from the production plant. Amount of discharged

S. Palamutcu (✉)
Pamukkale University, Denizli, Turkey
e-mail: spalamut@pau.edu.tr

S.S. Muthu (ed.), *Textiles and Clothing Sustainability*, Textile Science and Clothing Technology, DOI 10.1007/978-981-10-2474-0_1

water depends on the selected process; chemical auxiliary load depends on the type of dyeing and finishing chemistry types; and temperature of the discharged water depends on both selected processing method and energy exchanger installation existence of the plant. Another wastewater and energy-consuming issue of a textile item is confronted during the home laundry period of consumer, where washing, drying, and ironing processes leave big size of footprint. This chapter involves with the latest textile production technologies that improve sustainability feature of a textile product.

Keywords Sustainable · Environmental footprint · Textile production · Laundry · Consumer · Biodegradable · Toxic · Energy consuming

1 Introduction

Textile manufacturing process steps of fiber growing-manufacturing, spinning, weaving, knitting, nonwoven, wet processing, and ready-made product manufacturing are all cause some different level of environmental effect.

Textile products are manufactured using fiber which is the key material to define processing technology, application area, lifetime, and sustainability feature of a textile item. Textile products and textile sector itself are inevitable sector in our daily life that is continuously enlarging from medical applications to the civil engineering applications and nanofabrication technologies to the space technologies. Along the increasing textile product consumption in the developed and developing countries of the world, fiber demand on the sector has also rising that has reached about 95,6 million tons in 2015 (Fig. 1).

Share of oil-based fiber groups is the biggest with 62.1 %. Cotton fiber has about 25.2 % share, wood-based regenerated fiber groups have about 6.4 % share, cellulosic and protein-based other fiber groups have about 5.1 % share, and wool has

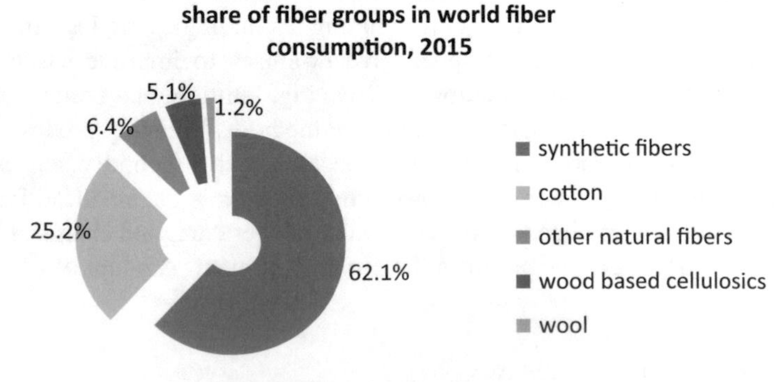

Fig. 1 Share of fiber groups in world fiber consumption in 2015 (Lenzing.com)

about 1.2 % share of whole fiber consumption. Fiber growing or manufacturing conditions and used technologies are the key factors to understand which fiber groups have better sustainability features. Natural, synthetic, and regenerated fiber groups have different advantages and disadvantages to be classified as the most sustainable fiber groups.

When cotton and polyester fibers, which represent the most commonly used fibers of natural and synthetic fiber groups orderly, are compared from the view point of sustainability features, a big confusion comes to stage. Cotton is natural and natural has so far become synonymous with green, clean, and even sustainable. However, it is not entirely acceptable. Pesticide and insecticides that are used during cotton growing processes cause high load of toxicological effect on the environment. Another important environmental influence of cotton growing is huge amount of irrigation water consumption. Concerning to overcome such environmental influences, organic cotton use is flourishing where intensive chemical use is abandoned, but it is still water-intensive. As the highest consumed synthetic fiber type, polyester fiber requires less water comparing to cotton fiber growing. However, it is more energy-intensive requiring oil to produce, thereby contributing to global warming as a result of harmful greenhouse gas emission.

Yarn manufacturing is the second main processing stage of a conventional textile item. Yarn preparation stages of yarn manufacturing have almost been the same in principle since last two century. Technologic improvements have been focused on high production speed, less waste, high efficiency, and less labor. Yarn spinning machinery and their technologic improvements are though quite revolutionary comparing to preparation stage machinery. Ring-spinning machinery and its technologic improvements have only focused on production speed and yarn quality, and principle technology has not changed. Besides conventional ring-spinning technology, new yarn spinning machinery and principles are introduced and implemented into the yarn manufacturing industry. Rotor spinning, friction spinning, air vortex spinning, and air-jet spinning are new generation of spinning machineries where high-speed yarn productions have become possible. Besides high-speed production, new spinning machineries require shorter prespinning and postspinning processes and machineries that result better sustainability features of spinning technology.

After production of yarn, next operation is the processing of textile surface via conventional weaving, knitting, or braiding technologies. Each three commonly used textile surface production technologies have been practically used techniques for a few thousand years. Ancient findings verify that weft and warp used weaving principle, looped knitting principle, and knotted braiding principle that had been used as handcraft techniques even before practical production of yarn spinning.

Woven surface production requires more number of preparation stages than knitting and braiding processes. Preweaving and postweaving processes of sizing and desizing processes are high energy consuming and wastewater discharging production steps that cause increased environmental effect. Weaving machinery itself has also had environmental effect which is continually improved to provide better production efficiency with high-speed production.

Wet processing of textile products causes high level of environmental influence as a result of intensive energy consumption and chemical material-loaded wastewater discharge. Wet processes are categorized as prefinishing, dyeing, printing, and finishing where each has its own characteristic environmental influence. Consumer using period of a textile product also causes high amount of environmental effect as a result of washing, drying, and ironing habits of the consumer. Such consumer habits are started to managed with social responsibility campaigns and promotions. As technological improvements about environmental effect of consumer behavior, washing machine adjustments, washing additives, and washing temperatures are targeted for any possible development.

2 History of Textile Technology

Textile products are essential aspects of our daily life as it was in the early history of human kind. Over the course of history, people of various cultures began to produce and wear clothing—not only as a means of protection, but also as a form of decoration and social distinction. Textile clothing has, in a sense, become exchangeable "second skin" of human (Deutsches museum). Historical findings have shown that very first textile surfaces have been used in the Europe—in the land of current Czech Republic—about 29,000–24,000 years ago (Parks 2000). Very first textile fibrous materials that have been used to make textile surfaces are wild plants of flax, hemp, and nettles. Other evidences of early weaving come from print and fragments of fibers found in Ukraine, Moldova, and Central France from 15,000 BC (Fig. 2), Asia, North America, South America, the Near East, and Switzerland are dated between 11,000 and 8000 BC. An archeological finding of 18,000 BC is a bone needle with eye, which may be used for joining fabric-like surfaces together. With the beginning of Neolithic Revolution after 8000 BC, domestication of sheep and use of clay tablets from the Near East, and first hand looms that make weaving, are started to be used (etn-net). Woven and intertwined textiles made of flax fibers, rug findings, and other findings of a warp-weighted loom have been found in the Çatalhöyük, Anatolia—Turkish Republic, and belong to 7000–6000 BC. During the Neolithic Era—between 5000 and 2500 BC—sheep breeding has been widened all over to Mediterranean area from the Black Sea along the valleys of Danube and Rhine up to present Belgium. Use of textile hemp fibers in Chinese Yang-Shao culture has been recorded. Wool and linen spinning and weaving in Egypt, Mesopotamia, Pakistan, and up to the Alps have been widened. And cultivation of silkworms in China has been started.

Between the age of 2000 and 700 BC, Bronze Age has just been started in Europe. In these times, textile materials are used for sail clothing, and cotton clothing has been started in Mesopotamia, in Babylon, and in India. Introduction to

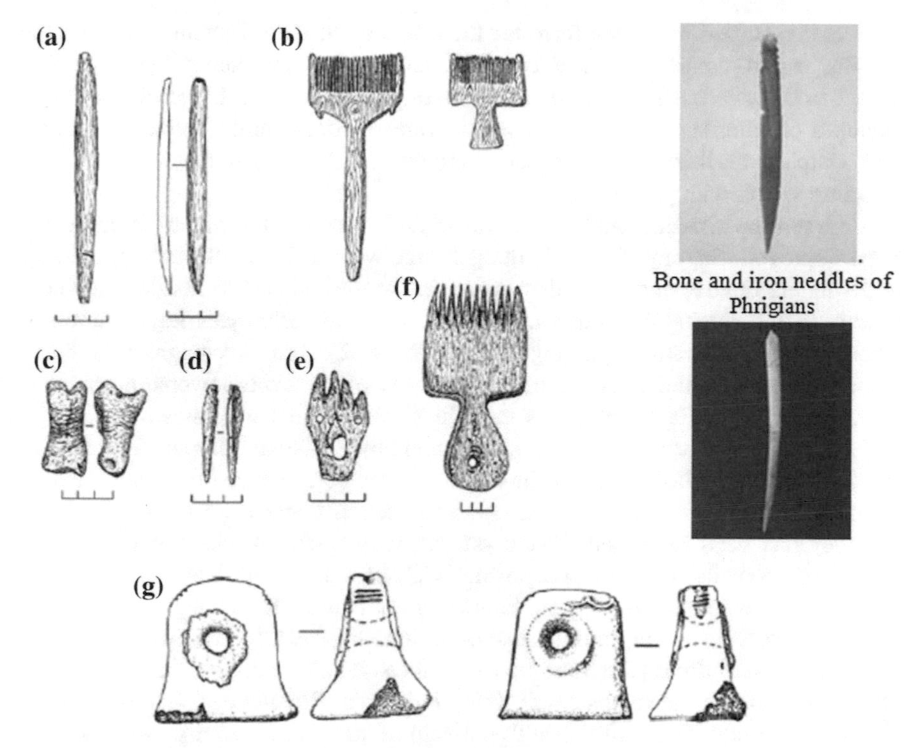

Fig. 2 Textile devices: **a**—curved awls from the Botai settlement; **b**—Slav combs for combing fibers; **c**—spool for winding threads from the Liman settlement, Ukraine; **d**—needle from the Liman settlement, Ukraine; **e**—comb from the Arich Burial Ground, Armenia; **f**—comb from the Uzerliktepe settlement; **g**—cone loom weights from the Galyugai settlement, northern Caucasus (**a** and **c**–**f** bone; **b** wood; **g** lay) (Shishlina et al. 2000)

knitting has been presumed. A cotton plantation in Assyria has been started. First indications of early Silk Road have started to be established.

In antiquity age, between 500 and 600 AD, cotton plantation and silkworm cultivation have been widened. Threaded loom and draw loom are introduced in China. Hand-spinning wheel is used in Asia. Knitted clothing has started to be seen in Upper Egypt. The knowledge on silkworm breeding arrived in Byzantium.

In the middle age, between 600 and 1492, sheep has started to be mentioned in England. Arabs conquer the southern part of the Mediterranean area, including Sicily and Spain. They bring along high-level textile culture of sheep breeding, cotton plantation, silk production, and the knitting technique from Egypt and the Near East. The medieval industrial revolution has started in Europe. Wool dyers' guilds are established in Germany. The "Hindustan Wheel," a hand-spinning wheel for cotton (presumably from India), is presented in Paris and further adapted for wool, doubling the output. The "filatoio," a silk reeling, twisting, and winding device, is recorded in written form for the first time in Italy. England becomes the leading export country for woolen cloth. Continuous producing flyer spinning wheel is discovered in 1480. In the "Codex Atlanticus" of Leonardo da Vinci sketches of spinning machines, automatic yarn dividers, yarn winding machines, and warping machines are published; around 1497, he had developed a flyer spinning wheel with yarn divider.

Between the sixteenth and eighteenth centuries, textile machinery industry has been improved. Saxony Wheel, knitting frame, water-driven silk reeling, twisting and winding device, and water-driven engines are introduced to textile machinery sector. Indigo dyeing has been introduced. Printing techniques have started to improve. "Water frame Spinning Machine" by Richard Arkwright was being patented. Patent on the invention of a "Power Loom," a steam-powered mechanical loom by Edmund Cartwright, was used in weaving mills with steam engines in 1787. Invention of printing from wooden rollers by Scotsman Thomas Bell; help to increase the production speed of the textile printer upto the 20 times more. Patent on the "Cotton Gin" machine to separate the cotton fibers from the seeds by Eli Whitney has been registered. Patent on pattern weaving machine with punched cards has been registered by Joseph-Marie Charles Jacquard. Cellulose has been gained from wood by A. Payen, France. Introduction of sewing machine with improved work-advancing motion has been done by Isaak Merrit Singer, USA.

In the modern times after 1915 up to our days, textile sector has influenced by a few revolutionary movements of "First Industrial Revolution" introduction of mechanical production facilities with the help of water and steam power, "Second Industrial Revolution" introduction of division of labor and mass production with the help of electrical energy, and "Third Industrial Revolution" use of electronic and IT systems that further automate production. All three industrial revolutions have positive influences on production speed, efficiency, and quality improvements in textile sector.

And the latest industry revolution called Industry 4.0 connects embedded system production technologies and smart production processes to pave the way for a new technological age (gtai.de). Influence of the latest industrial revolution on textile sector will probably be in the quality improvement and efficiency increase that result more textile production and more environmental effect.

Increased textile consumption and resulting environmental side effects make some part of the society alerted about sustainability approach in textile in 1990s. As a result of industrial development, the urge of societies and craving of textile consumption have started to show their side effects on the planet and some consumers have started to change their behavior (Wicker and Becken 2013). Consumers start to question operations of companies and make their purchasing decisions more consciously (Galbreth and Ghosh 2013). Purchasing power of consumers has become a controlling instrument over the companies, pushing them to develop more sustainable approaches and use more sustainable production technologies.

3 Sustainable Technology and Textile Manufacturing

Sustainable development is a development that meets the needs of the present without compromising the ability of future generations to meet their own needs (Mulder et al. 2011). Sustainable development eventuates on the social, cultural, economic, and technological bases where the last one is one inevitable asset.

Modern technology improvements bring two features: solving the current problems and creating some new problems. It means every new technology brings its side effects along its advantages. Sustainable development is the mission of the society, and it is closely linked to the use of technology. Without technology improvement, society would have risk of collapse. To maintain balance between technology for improving society and technology for sustainability has now become an asset. In the modern societies, technological improvements should be established concerning the equal priority of "people," "profit," and "planet" to merge in the sustainable technology frame.

In today's world, technology has already become globalized and some specific products are only accessible via limited number of enterprises. Responsibility of those technology designers or developers becomes even greater. Any new and different touch or approach on a new development or design has tremendous effects on various aspects of the societies, in the short term or in the long term.

In textile processes and products, responsibility of technology developers and designers are even more pervasive. Any single initiation for a technology change on fiber production, yarn production, or chemical application procedure directly influences millions of people, soil, and water all around the world.

Textile and clothing industry is one of the most widened sectors in the world. Every single person participates in some part of the textile sector in either the production, manufacturing, or consumer parts. Textile manufacturing chain starts with fiber growing of natural fiber and manufacturing of man-made fiber. Consumption of fiber-based products has been increasing as a result of population growth of emerging markets, rising life standards, age structure in the developing countries, and increase of textile-based product types and amount of their consumption (Muthu 2015). Every year, fiber demand has increased at changing rates between 1.5 and 3.0 % correlating to increased incomes in the society.

Every gram of fiber submission to the market triggers the textile production chain of yarn, fabric, wet processes, and garment manufacturing. Besides the desired textile production activities, undesired side effects of these activities have also occurred and along environmental harms are erected on variety of degrees.

Manufacturing technologies that are utilized during the production phases of a textile item differs depending of the product properties. As it is mentioned in Fig. 3, energy profile and consecutively environmental load of different type of textile item (Tshirt, blouse, and carpet) have different energy profiles depending on the raw material and selected manufacturing techniques of the textile item, and consumer habits. Estimated total energy consumption is approximately 109 MJ per T-shirt, 51 MJ per blouse, and 390 MJ/m^2 of carpet. When the given data are analyzed

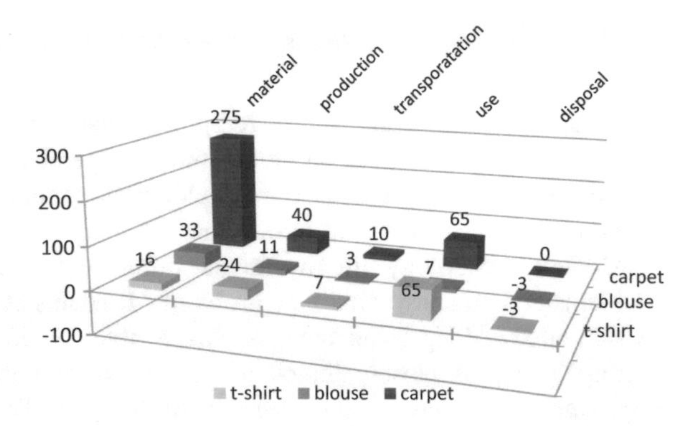

Fig. 3 Primary energy profile for piece of T-shirt and blouse and square meter of carpet, MJ (Muthu 2015)

from the view of sustainable technology approach, material and production phases should be involved, and any achieved decrease on these phases should be accepted as sustainable technology improvements. For the carpet, the material production phase represents approximately 71 % of the total energy. This is partly the result of the relatively large energy consumption in the production of the synthetic fiber polyamide—approximately 160 MJ/kg—compared to about 50 MJ/kg for cotton. Any energy profile decrease of fiber manufacturing brings significant achievements on the sustainability degree.

As similar consumer items, energy consumption rate of the viscose blouse and cotton T-shirt is not the same. The material phase energy consumption rates of the blouse and T-shirt are about 65 and 15 % of the total energy consecutively.

Viscose fiber manufacturing of blouse requires more energy than cotton fiber growing. The breakdown of energy for piece of viscose blouse is strikingly different compared to that of the T-shirt. The production phase energy requirement of blouse is 11 MJ that is less than 24 MJ for T-shirt. Production phase of blouse and T-shirt includes yarn spinning, knitting/weaving, and wet processing. To compare the yarn spinning phase, unit cotton yarn production requires more energy than viscose yarn production; weaving processes of blouse requires more energy than knitting process of T-shirt; and wet-processing phases of blouse and T-shirt have similar rates of energy. In total energy consumption on production, T-shirt production requires more energy than those of blouse. From the view of sustainable textile production technology approach, viscose blouse and cotton T-shirt can be classified in the same group.

4 Sustainable Fiber Technologies

Fiber is the basic raw material of textile products. Natural, regenerated, and synthetic-based fiber groups can be redefined, and their growing, processing, and manufacturing stages can be redesigned. New developed, regenerated, and synthetic fiber groups are mostly designed in the frame of sustainable technology as much as possible. Sustainable textile fiber technology development and design concept have been flourished after 1990s, and prosperous sustainable fiber types are introduced to the market. Besides the introduction of new generic fibers, the conventionally consumed fibers are limited to the cotton, polyester, cellulosic regenerated fiber, nylon, polypropylene, and wool fiber. In the research about life cycle impact assessment (LCIA), woven textile product of cotton, polyester, nylon, acryl, and elastane fiber is evaluated using the LCIA method (Velden et al. 2014). In their work, unit eco-cost of fiber types in (euro/kg) is estimated. Eco-cost is a measure to express the amount of environmental burden of a product on the basis of prevention of that burden. It has reported that from raw material extraction to manufactured textile, acryl and polyester have the least impact on the environment (followed by elastane and nylon) and cotton represents the highest environmental burden (Vogtländer 2013).

Organic cotton growing and processing technologies are one good example of natural fiber group in the sustainable fiber technology concept. In the organic cotton growing, only environmentally low-impact methods and materials are used. Organic production systems replenish and maintain soil fertility, reduce the use of toxic and persistent pesticides and fertilizers, and build biologically diverse agriculture. In the frame of sustainable technology, seed of the cotton, its required agricultural steps, and harvesting methods are designed concerning the environment. Besides organic cotton growing concept, there are other cotton growing programs that are designed and aimed to decrease the environmental deterioration level of cotton growing. Some of commonly known programs are Better Cotton Initiative (BCI) (bettercotton.org), cleaner cotton (sustainablecotton.org), and natural colored cotton (naturecolored.com).

Polyester staple and filament fiber have the highest amount—over 55 million tons—of market volume in the all type of fiber groups, and any possible sustainability approach in the polyester fiber processing technologies would be welcomed by environmental activist. Market share of polyester fiber has reached to the share of cotton for the first time in year of 2000, and later on, increasing trend has continued. Every single person in the world has already got in touch with a polyester fiber. Production of polyester fiber is based on petrochemicals, which are non-biodegradable and inherently unsustainable, and processing stages consume large amounts of water and energy.

On the contrary, yarn manufacturing and surface manufacturing stages of polyester fiber have less influence on the nature comparing to the those stages of cotton fiber. Additionally, consumer using time influences of washing, drying, and ironing processes causes lower environmental influences.

Recycled polyester fiber is another new feature of polyester fiber that is also acceptable as inherently unsustainable. Recycling process reclaims the non-biodegradable and inherently unsustainable properties of polyester fiber, and only environmentally favorable properties of polyester fiber stay with the recycled polyester fiber.

Regenerated biodegradable fiber types are introduced to the textile markets favoring their natural raw material backgrounds. They are produced from renewable cellulosic plants such as beech trees, pine trees, and bamboo. The most known regenerated fiber production technologies are viscose rayon—the first-generation technology, modal—the second-generation technology, and lyocell—the third-generation technology. Regenerated fiber production technologies vary depending on their processing steps, chemical intensities, environmentally friendliness level, and resulted fiber properties of luster, softness, drape, absorbency characteristics, and some other functional properties. Bamboo and Tencel® fibers are also market-oriented regenerated fiber types where their raw materials are also cellulosic plants. However, it should be stated again that much of the total environmental impact of textile items comes from their consumer phase of house care. Among the regenerated cellulosic fiber types, lyocell fiber is the most environmentally friendly fiber where it does not require fabric softener or whitening agents, and energy–water consumption can be decreased due to shorter washing machine cycles.

Polylactic acid fiber (PLA) is another manufactured fiber known as corn fiber and is a synthetic fiber made of polylactic acid or ester extracted from natural sugar of cereals (mainly corns) and beets through solution spinning or melt spinning. As a melt-spinnable fiber with a vegetable source, PLA has many of the advantages of both synthetic and natural fibers. Perhaps most distinctive among these, though, is the fact that, like natural fibers, its raw material is renewable, non-polluting, and compostable. PLA is less environmentally costly than polymers that are recyclable, because there is a limit to the number of recycling iterations that can occur before the material loses its usefulness. PLA is even less environmentally costly than other biodegradable thermoplastics, since the entire mass of PLA can eventually be reconverted into new PLA, whereas many other biodegradable thermoplastics incorporate at least some material derived from fossil fuels. PLA is not a perfectly sustainable polymer, since some energy must be irretrievably used in its polymerization and in converting the polymer into fibers and fabrics. But it offers superior sustainability and lower environmental impact than any other non-cellulosic synthetic fiber and possibly even superior to some natural fibers (Dugan 2001). Fiber has the desired merits of biodegradable and environmentally friendly properties (Xiong and Li 2010).

Fiber properties of PLA can be planned through its fundamental polymer chemistry processing stages, and PLA fiber becomes a suitable fiber for a wide range of textile applications. The mostly known fiber properties are low moisture absorption and high wicking, low flammability and smoke generation, high resistance to ultra violet (UV) light, a low index of refraction, and lower specific gravity. As one biodegradable fiber PLA, its biodegradation degree depends on its molecular weight, crystallinity, geometry, temperature, moisture pH, and the presence of microorganism in the surrounding area (Ju et al. 2015). PLA textile fiber resembles polyethylene terephthalate (PET) fiber in some fiber properties of higher tenacity than natural fibers, with excellent moisture transport away from the skin surface (Hagen 2013). It has been stated in the report of Hagen that global warming potential of PLA fiber is lower than that of PET, cotton, and polypropylene PP fiber, and higher than tencel, Lenzing Modal, and Lenzing Viscose. Water consumption of PLA fiber is lower than PET, PP, regenerated cellulosic fibers, and cotton and other natural cellulosic fiber.

5 Sustainable Yarn Production Technologies

Yarn spinning process is the second step of conventional textile production route. Evaluation factors about the sustainability level of spinning technologies are machinery lines, machine working speeds, fiber type, consumed energy type, yarn thickness, twist level, and yarn package type. Energy consumption per unit yarn production is the most influential parameter to understand the sustainability level of the yarn spinning process.

Technologically, yarn production has two main groups: staple yarn types and continuous filament yarn types, which require different groups of machinery with their different technologies. Staple yarn spinning processes are mostly involved with fiber separations, ventilation, and guidance which require machineries with mechanical enforcement and aerodynamic systems. Texturing machinery is mostly involved with heating, cooling units, mechanical movements, and aerodynamic enforcement systems. Both staple and texturing yarn production processing steps require air-conditioning, compressors, lighting, and cleaning appliances.

Spinning machinery lines of ring spinning, rotor spinning, and air-jet spinning and the complementary installations consume electric energy (Koç and Kaplan 2007; Kaplan and Koç 2010). Complementary installations are accepted as constant energy consumption where it does not change with the other production parameters of yarn number, twist level, and raw material properties.

Ring-spinning technology starts with the invention of the spinning mule, or mule jenny, in 1779 by Samuel Crompton. Since then, the principle of ring spinning has not changed; however, the textile machinery industry has worked to develop faster and more reliable and flexible spinning technologies.

Ring-spinning machinery is an electric energy-intensive process, which consumes about 72 % of the total monthly energy consumption of an average spinning plant, where air-conditioning comprises 16 % of the total energy consumption (Koç and Kaplan 2007). And the rest of the electric energy consumption is comprised by other processes. Ring-spinning machine and its mechanisms consume high amount of energy to manage spinning of fiber into the yarn. Energy consumption share of the machine is not constant, and it changes depending on the yarn count. The amount of energy consumption for different types and counts of yarn varies; finer yarn needs more energy for all types of yarn; higher twist level of yarns for the same yarn number requires more energy, and combed yarns require more energy than the same count and twist level of carded yarns.

From the view of sustainability frame, ring-spinning technology and ring spun yarns have the highest level of environmental influence and the ring-spinning process adds the highest level of environmental load on to the textile item; even the yarn is spun using natural fiber or biodegradable fiber. Any improvement in sustainability degree of ring spun yarns can become possible with the increase of spinning speed, which is a challenge, decrease of energy consumption with efficient engines and driving systems, appropriate oiling, lightweight bobbin use, and development on the material properties of ring traveler.

Open-end rotor yarn technology is about 50 years old and is the second commonly accepted spinning technology in the world. Production machinery line of the rotor spinning line is shorter than the machinery line of ring-spinning line. Yarn production speed of the rotor spinning machine is about 5–10 times higher than ring spun yarn.

Open-end rotor spinning machine itself is also electric energy-intensive machine that consumes about 78.4 % of the total monthly energy consumption of an average spinning plant, where air-conditioning comprises 16 % of the total energy consumption (Kaplan and Koç 2010). Energy consumption share of the rotor spinning machine is not constant, and it changes depending on the yarn count, twist level, fiber type, and properties of used spinning elements. The amount of energy consumption for finer yarn needs more energy than all types of yarn; higher twist level of yarns for the same yarn number requires more energy.

The current rotor spinning technology has been consistently developed in the direction of higher productivity and lower energy consumption. The improved spinning technology allows the same yarn quality to be produced with less twist and smaller rotors. Such approaches help to increase productivity and save energy (bluecompetence.net). Furthermore, the driving components of the rotor spinning have been optimized, and more spinning positions with lower individual energy consumption are concentrated with minimum space requirements in single machine. As a result, it has become possible to decrease energy consumption per kilogram of yarn about 25 % without any quality deterioration on the spun yarn.

From the view of sustainability frame, rotor spinning technology and rotor spun yarns have lower level of environmental influence than ring spun yarns. Any improvement in sustainability degree of rotor spun yarns can become possible with the further increase of spinning speed, which is a challenge, efficient engines and driving systems, improved and engineered surface properties on the spinning elements of rotor, navel, opening cylinder, and other spinning elements.

Air-jet yarn spinning technology is the newest spinning technology that has been accepted by the spinning markets in the last 20 years. Production machinery line of the air-jet spinning line is shorter than the machinery line of ring-spinning machinery line.

The air-jet system virtually totally integrates the fibers into the yarn body and completes twist insertion into the yarn. Twist insertion and fiber integration onto the yarn strand are completed only when the air and no mechanical parts are directly involved in the twist insertion process. It has high yarn spinning rate up to 450 m/min, where normal ring-spinning delivery speeds are 15–27 m/min, and rotor spinning speeds are 130–250 m/min (textileworld.com) (Fig. 4).

However, the main drawback regarding the technology is the high energy consumption due to the compressed air usage. Since the cost of energy is an important and challenging issue, especially in some countries, such advantageous factor of spinning technology may become a limiting factor for the selection of the air-jet spinning technology. Despite of high energy consumption of the air-jet spinning machine, it has been reported that future trend in air-jet yarn production is

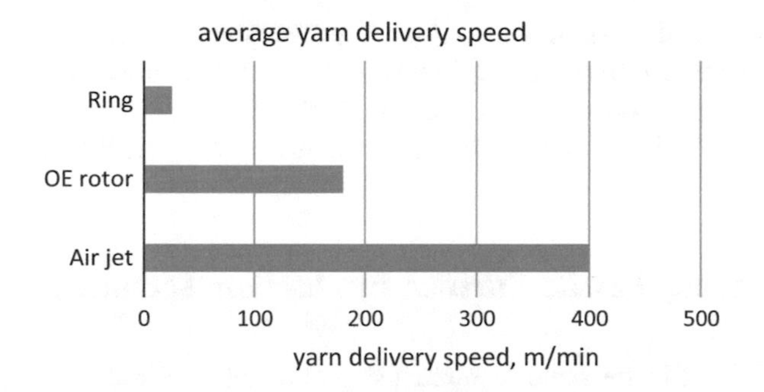

Fig. 4 Average yarn delivery speed levels of three main spinning technologies

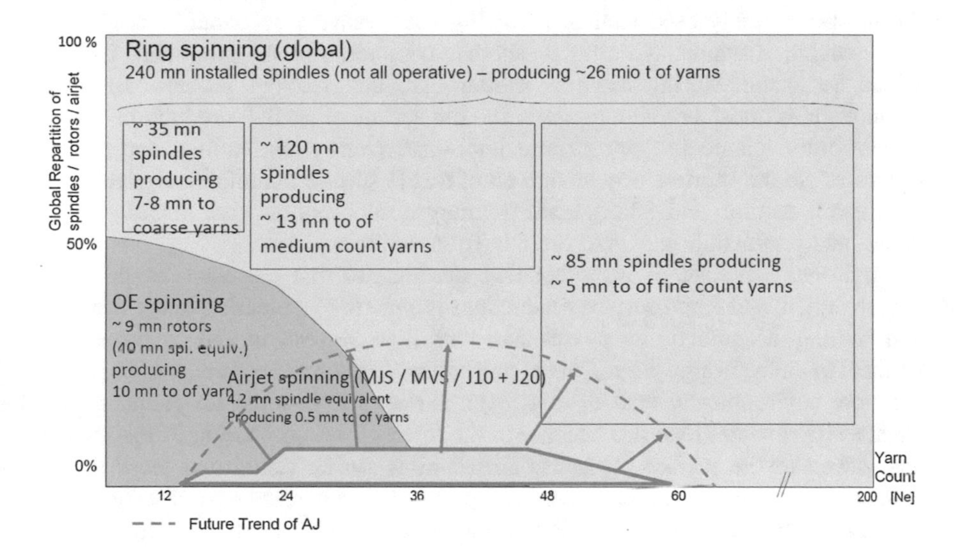

Fig. 5 Installed spindles and short staple yarn production in 2010 (https://www.unido.org/fileadmin/user_media/UNIDO_Worldwide/Offices/UNIDO_Offices/South_Africa/Report_Phase_1_and_2_XPRAF08005.pdf)

on the way of more expansion (https://www.unido.org/fileadmin/user_media/UNIDO_Worldwide/Offices/UNIDO_Offices/South_Africa/Report_Phase_1_and_2_XPRAF08005.pdf) over the market share of rotor yarn and ring yarn (Fig. 5).

Currently in the world, 26 million tons of ring spun, 9 million tons of OE rotor spun, and 05 million tons of air-jet spun yarn manufacturing are realized annually. It has been estimated by Gherzi that amount of air-jet spun yarn will increase annual manufacturing of 2 million tons until 2020.

Space needs for the modern air-jet machines are 25 % less than that for ring-spinning equipment producing the same capacity, thereby reducing building costs. Also, the smaller area requires less climate control, resulting in further substantial savings. From the view of sustainability frame, air-jet spinning technology and air-jet spun yarns have the lowest level of environmental influence among rotor and ring spun yarns. Any improvement in sustainability degree of air-jet spun yarns can become possible with the further increase of spinning speed, decrease of energy consumption, utilization of efficient engines and driving systems, and improved and engineered air-jet units on the spinning machine.

6 Sustainable Textile Surface Production Technologies

Textile surfaces can be manufactured using different methods and technologies. Weaving, knitting, braiding, nonwoven, tufting, and felting technologies are commonly used textile surface manufacturing technologies. Weaving technology is the most commonly used technic involving with weaving preparation machinery and weaving machine. Knitting is another common surface producing technic where the manufacturing stage is shorter than the weaving process. Braiding technology is most commonly used for the technical textile production where sustainability feature is getting more important. Nonwoven, tufting, and felting processes are the shortest way among all of textile surface manufacturing technics. Nonwoven, tufting, and felting manufacturing do not require yarn manufacturing phase, and production may start directly from the fiber phase.

Environmental load of manufacturing machineries that are used for the nonwoven, tufted, and felted surface productions is quite low comparing to the weaving and knitting. Manufacturing phases start with fiber processing and continue with surface manufacturing phase, yarn processing machinesries are not required; therefore environmental load of nonwoven surface has become lower inherently. Total environmental load has become to the lowest level comparing to the knitted and woven textile surface products. Therefore, it can be stated that those textile surfaces that are manufactured directly from the fiber to surface have the high level of sustainability feature; only limitation might occur with the use of unsustainable fiber utilizations on the surface production.

Weaving preparation and weaving technologies employ preparation machines of warping, sizing, and weft preparation; weaving machine; and postweaving process desizing process. All process technologies require high amount of electric and heat energies. Warping machine and weft preparation machine consume only electric energy with reasonable level of environmental load, whereas sizing and desizing processes consume high level of heat energy beside electricity and result high degree of environmental load.

Warp and weft preparation machineries are electric energy-consuming systems, and their energy efficiency might be improved with controlled maintenance, cleaning, efficient engines and driving systems, and other mechanical equipment on the machines. Achievements of waste yarn decrease on these processes result in improvements in the sustainability degree of the weaving preparation processes. Sizing process of weaving preparation phases is the most important process from the view of sustainable weaving technology. Reduction of heat energy consumption and environmental load of the sizing auxiliary are two key factors to improve sustainability feature of the sizing process. Reduced steam temperature utilization and cold sizing agent use are offered possibilities for decrease of environmental load of the sizing process. Besides the continuous improvement attempts on the environmental load of sizing process, cutting off the whole process via utilization of high strength of warp yarns is another revolutionary and challenging option. To achieve efficient size-free weaving, superior quality of warp yarn is required. The highest possible uniformity and consistency and the least amount of hairiness are significantly important yarn properties, which are the function of the selected fiber quality, yarn structure, and spinning parameters (Sawhney 2006).

After the completion of the weaving process, size of the fabric needs to be removed to ensure the quality of consecutive finishing processes. Desizing process consumes large amount of discharged water, which is about 40–50 % of the total wastewater release of the plant. Most of the size removed during desizing is discharged in effluent form and causes high level of environmental load.

Weaving machine technologies vary depending on their weft insertion technics such as shuttle looms and shuttle-less looms. On the basis of weft insertion mechanism, shuttle-less looms are further classified into projectile, rapier, water-jet, and air-jet looms.

Air-jet looms, which use the newest weft insertion technology, have become popular in the textile industry due to their high productivity and better control features. Production speed of shuttle looms and shuttle-less looms of projectile, rapier, water-jet looms has some disadvantages comparing to the air-jet looms.

High production speed and low number of worker need are two main advantages of air-jet looms. The air-jet weaving machine combines high performance with low manufacturing requirements, because differently from rapier and projectile machines, the filling medium is just air and no mechanical parts are directly involved in the weft insertion process. It has high production rate up to 1100 weft insertions per minute, and it covers a wide range of processing yarns such as spun and continuous filament yarns. However, the main drawback regarding the technology is the very high energy consumption due to the compressed air usage which is required during the weft insertion process. Since the cost of energy has a systematic increasing trend, power consumption is still a challenging issue. In particular, it is the limiting factor for such technology in the countries, where energy costs represent a large share of the manufacturing (Grassi et al. 2016). So far reduction of compressed air consumption without compromising the performance of air-jet weaving machine and quality of woven fabric is one major study area among researchers and technology developers.

Any possible decrease of energy consumption per each weft insertion value on the weaving machine will provide improvement in the sustainability level of weaving together with weaving preparation phases of warping, sizing–desizing, and weft preparation processes.

Knitting technology is the second commonly used textile surface manufacturing technique that has been practiced for the thousand years as handcraft technique. In modern textile technology concept, knitting textile surface is formed by interlocking or intermeshing loops of single yarn, set of yarns, or more group of yarns. Knitting process is performed using two principle technologies of weft and warp knitting, where several different types of machines are used. Knitting machinery consumes electric energy in their driving systems and control systems. Consumption of energy and related environmental load of the knitting process are quite low comparing to those of woven textile fabric. Any energy-intensive special machinery or equipment does not require before and after knitting process. From the view of sustainability frame, knitting technology has better sustainability feature than woven fabrics.

Seamless knitting technology is a new technology. Seamless knitting technology has entered the mainstream in the knitwear market at the end of last century. Seamless entire garment knitting was first introduced in 1995, since then it has spread over the different group of clothing products from underwear to sport clothing. Seamless knitting technology requires minimal or no cutting and sewing process to complete an entire garment. This innovative technology promises elimination of postlabor work, more comfort and better fit to the consumer by eliminating seams. And knitting technology itself offers lower energy consumption and less environmental load as heritage of knitting technology. Advantages of the seamless knitting production technology can be listed as saving on labor, time, cost, and environmental load where the lowest possible level of waste disposal, reduced need of fiber and yarn.

7 Sustainable Wet-Processing Technologies

Wet processing on textile manufacturing includes pretreatment of fabric before dyeing or printing process, dyeing–coloration–printing processes, and postprocessing or finishing processes. All stages of wet processing require high amount of energy; water is required for the manufacturing stages of washing, heating, drying, and other wet-processing steps. As a result of high energy and water consumption, wastewater and emission loads of the processes are also high in the traditional textile wet-processing plants.

In conventional textile dyeing, water has double effect on the environment, as consumption of freshwater and its lateral discharge. On average, conventional textile wet processing 100–150 L of water is required to process 1 kg of textile material. In the world about 28 billion kilos of textiles, material is being processed annually in the water as solvent medium.

There are many researches, projects, initiations, and programs to make contribution to the environmental influences of desizing, bleaching, scouring, dyeing, washing, drying, finishing, and printing machineries, principles, and technologies. Chemical auxiliaries that are intensively consumed in the wet-processing stages are also studied to make improvement in their environmental influences related to application techniques, content in the wastewater, effects to the consumer, home laundering practices, and biodegradation features and harms to the natural life.

Among the high number of scientific, engineering, social, and cultural contented studies, it has been stated that sustainable textile wet processing becomes possible by using intensive effluent treatment in the plants, utilization of eco-friendly dyes and chemicals, and new approach of application technologies. New technologies are expected to focus on the reduction in energy consumption, elimination of process water and wastewater discharge, and decrease of chemicals that would be accepted as real breakthrough actions in textile wet-processing industry.

Supercritical dyeing technology, as one emerging process for dyeing, uses supercritical carbon dioxide as dyeing medium. Dyeing process is performed in a high-pressure vessel called an autoclave. Carbon dioxide exists as a supercritical fluid at temperatures at about 31 °C and pressures above 72 bar. The anhydrous process offers a number of ecological and economic advantages such as no preparation of processing water and low energy consumption for heating up the liquor (Hasanbeigi 2010). Hence, the elimination of process water and auxiliary consumption opens a great frame of opportunities for the textile dyeing industry, and industrial-sized commercial machineries are already introduced to the market as fruition of the supercritical dyeing technology by DyeCoo Textile Systems BV (dyecoo.com).

Ultrasonic assisted wet-processing technology is another emerging technology related to the wet-processing techniques. Ultrasonic assisted process is developed as an alternative method to conventional high-temperature processing technologies. Ultrasound equipment installed in the existing machines offers improved performance in fabric preparation and dyeing without impairing the properties of the processed materials. The advantages of the use of ultrasonic in textile wet-processing include energy savings due to reduced processing temperatures and time, and the lower consumption of auxiliary chemicals (Prince 2008).

The ultrasonic method has been effectively utilized in various fabric preparation processes including desizing, scouring, bleaching, and mercerization and auxiliary processes such as washing and laundering. Attempts have been made to analyze the effect of ultrasonic in dyeing processes on almost all types of fibers using direct reactive, acid, and disperse dyes. Ultrasonic waves accelerate the rate of diffusion of the dye inside the fiber with the enhanced wetting of fibers (Ramachandran et al. 2008). Industrial-sized commercial application of the process is also introduced to the market by DyeCoo Textile Systems BV.

AirDye method that is developed in California by Colorep (airdye.com) works with proprietary dyes that are heat transferred from paper to fabric in a one-step process. This can save between seven and 75 gal of water in the dying of a pound of fabric, save energy, and produces no harmful by-products. The technology uses 85 % less energy than traditional dyeing methods.

Plasma treatment in textile-finishing processes is another emerging technology that provides enhanced mechanical and electrical properties to the fabric, better wetting, dyeing, printing, and improved color fastness features of fabric.

Industrial-sized commercial application of plasma surface treatment is introduced to the market by Enercon Company (enerconind.com). Application of plasma treatment to textiles, filaments, and yarns is used for the primary purposes of desizing, dye uptake, printing ink adhesion, and final finishing treatments such as softening, hydrophilization, easy care, and anti-shrinkage. Enercon's Plasma™ atmospheric plasma surface treatment envelops and cleans the fibers of woven and nonwoven structures to enhance their wettability, printability, dye ability, and fiber capillarity.

Foam coating technology is offered as an alternative method for conventional wet-processing technologies. In the foam coating technology, instead of water, air is employed to make dilution of the treatment liquor and later textile application is conducted. The offered technology leads to saving in energy and water consumption of conventional wet-processing technologies. Saving can be reached up to the 80% reduction of water consumption and 60–65 % reduction in energy use, depending on the type of finishing treatment used. Besides savings on the water and energy, emission gases are minimized. Industrial-sized commercial application of foam coating is commonly used in different surface treatment processes for several decades, where textile application is also widening.

Digital printing technology is one revolutionary phenomenon in textile printing technology, and it has already managed to increase its application from apparel to home textiles increasingly taking market shares from screen printing. Worldwide known printing machine companies of Reggiani, Durst, Zimmer, Konica, Minolta, and MS have already designed and implemented their latest digital printing technologies to the textile printers allover the world. Utilization of digital printing technology instead of conventionally known printing technologies requires as little as 1 L of water per linear meter instead of the usual 50–100 L of water, saves more than 50 % energy, and reduces the CO_2 emissions up to 90 % (elvajet.com). The other offered advantages of digital textile printing are short-run printing orders, reduced chemical waste, widened printing size, reduced space utilization, reduced printing investment, quick sampling, flexible printing, and increased creative design choices.

Natural dye used in dyeing and printing processes is another option in the sustainable wet-processing technologies. It is anciently well-known technology, and it has become popular again with the increased awareness of sustainable products in the society. Worldwide known textile brands and department stores have started to prepare natural dyed and printed collections to imply their sensitivity to the nature. Research and commercialization works about the natural dyes are increased in the last decade that some natural dyestuff has already become classic and popular, which are indigo, madder, cochineal, weld, and cutch. These dyestuffs are researched to give almost any color with the exception of a few colors such as fluorescent and electric blues (dyes-pigments.com). Common application of natural dyes in the market can make the textile industry more competitive, by reducing production costs, eliminating expensive chemical auxiliary consumption, and decreasing the environmental load of dyeing chemicals.

8 Sustainable Technologies in Garment Manufacturing

Garment manufacturing or ultimate product manufacturing phase of textile manufacturing steps involves with cutting, sewing, ironing, and packaging processes in conventional textiles. **Cutting** process is mostly completed via electric engine drive cutting systems where low level of energy is consumed. The key factor for cutting process is the lowest possible cut waste ingeneration during the cutting of the fabric. It has been reported (Risanene 2005) that approximately 15 % of textiles intended for clothing ends up on the cutting room floor. To decrease the amount of cut waste of cutting phase, special computer-controlled patterning programs are developed and successfully used in the market.

Sewing process involves in different types of sewing machineries where reasonable amount of electric energy is consumed. Continues cleaning, oiling, and maintenance of electric engines and other appliances of the sewing machinery will help to manage the energy consumption on the sewing machineries.

Ironing process is the highest energy-consuming process in the garment manufacturing. Steam production and heating of iron are maintained using electric energy. In ironing process, worker's habits are the most influential parameter that might be improved to reduce energy consumption and improve sustainability degree of the ultimate product. From the view of technology, low energy-consuming machinery selection might be useful to improve sustainability level of the machinery.

Packaging materials and label tags of the manufactured garments have the highest level of environmental harm. All of the package material and label tags directly send to the waste lands right after reaching to their last destination at the consumer closed. Amount of package and label material, their type, and their biodegradation time are influential parameters. Types of used package materials are paper, nylon, polyester, or any other material.

9 Conclusion

Textile manufacturing technologies cause some high amount of environmental load on the nature as a result of their energy consumption, wastewater discharge, and other types of waste accumulation. All manufacturing operations and technologies create harmful influences on the nature; however, most of the processes are indispensable processes that are conventionally accepted. There is no possibility of zero impact, or entirely erasing of the environmental load, left behind a textile product. Only possible ways to reduce the environmental load are technology improvement, appropriate technology selection, and efficient technology management. Current technological developments offer several opportunities to take action of new production technology routes and models to build sustainable textile technology environment.

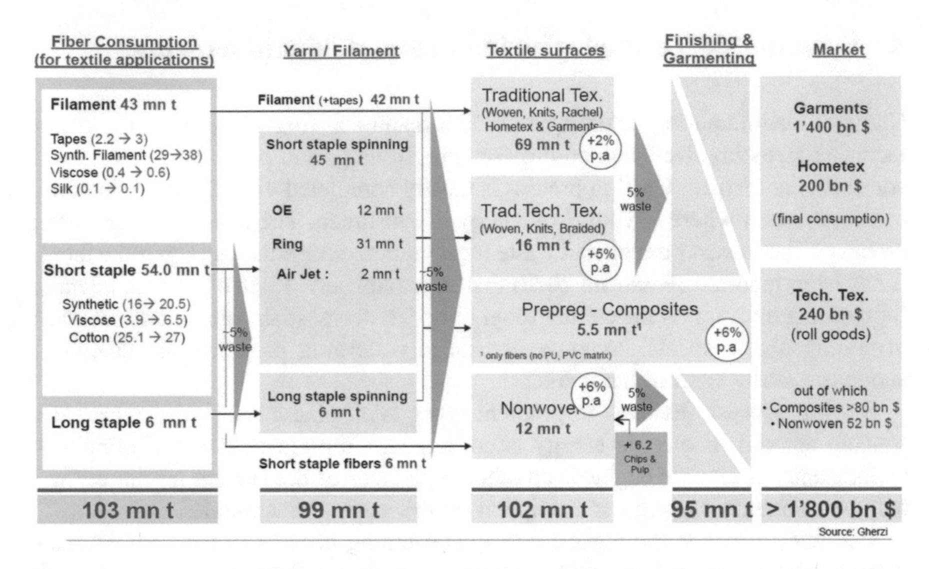

Fig. 6 Estimated worldwide textile added value chain in million ton, by Gherzi (itmf.org)

The estimated volume of the textile sector in the world is reported as 103 million ton of textile fiber submission and 1800 billion USD annual turnover for the year of 2020 in the ITMF report (Fig. 6).

To manage to increase the sustainability degree of such high amount of fiber-based textile products, some opportunities are offered by Gherzi: continuous specific R&D regarding sustainable fiber manufacturing or growing, new spinning processes or technological modification of known spinning technologies, new textile products of composite and nonwoven technologies, reinvention on the finishing machineries, chemicals and processes, and management of product groups and design trends based on local and traditional needs. Some other flourishing technologies are based on personalized production possibilities of body scanning, virtual personnel designer and styling via tablet, 3D printer utilization in textile, "Garment DNA" tracking from fiber to garmenting approach, payment through digital devices without paper receipt, and increased recycling–reusing and reducing approach.

Sustainable textile technology model is continuous improvement model where its main dynamics are technology developers, fashion designers, consumers, and manufacturers. Approach of technology developers and fashion designer is manageable that they would easily follow the main rules of sustainability. Consumer behaviors are not easily manageable concerning the sustainability; some portion of the consumer resists the certain rules of sustainability approach and products of sustainable technologies, whereas other portion of the consumers support the sustainability programs. In the sustainable textile technology utilization, manufacturers are another important player and they are dominated by small- and medium-sized enterprises (SMEs) where sustainable textile technology approaches are either not welcomed, not accepted, not well managed yet, or even not present.

References

Dugan, J. S. (2001). Novel properties of PLA fibers. *International Nonwoven Journal, 10*(3).
Galbreth, M. R., & Ghosh, B. (2013). Competition and sustainability: The impact of consumer awareness. *Decision Sciences* [e-journal], *44*(1), 127–159.
Grassi, C., Schroter, A., Gloy, Y. S., & Gries, T. (2016). Increasing the energy efficiency of air jet weaving based on a novel method to exploit energy savings potentials in production processes of the textile industry. *Journal of Environmental Science and Engineering B, 5*, 26–34. doi:10.17265/2162-5263/2016.01.004.
Hagen, R. (2013). The potential of PLA for the fiber market, Bioplastic Magazine 05/13, (Vol. 8, p. 12–15).
Hasanbeigi, A. (2010). Energy—Efficiency improvement opportunities for the textile industry. https://www.energystar.gov/sites/default/files/buildings/tools/EE_Guidebook_for_Textile_industry.pdf
Ju, D., Han, L., Guo, Z., Bian, J., Li, F., Chen, S., et al. (2015). Effect of diameter of poly (lactic acid) fiber on the physical properties of poly (ε-caprolactone). *International Journal of Biological Macromolecules, 76*, 49–57.
Kaplan, E., & Koç, E. (2010). Investigation of energy consumption in yarn production with special reference to open-end rotor spinning. *FIBRES & TEXTILES in Eastern Europe, 18* 2(79), 7–13.
Koç, E., & Kaplan, E. (2007). An investigation on energy consumption in yarn production with special reference to ring spinning. *FIBRES & TEXTILES in Eastern Europe, 15* 4(63).
Mulder, K., Ferrer, D., & Lente, H. (2011). *What is sustainable technology? Perceptions, paradoxes and possibilities*. England: Greenleaf Publishing Limited. 978-1-906093-50-1. https://www.greenleaf-publishing.com/technology
Muthu, S. (2015). *Handbook of life cycle assessments (LCA) of textile and clothing*. Woodhead Publishing.
Parks, L. (2000). *Scientific American discovering archeology* (pp. 26–28).
Prince, A. (2008). Energy conservation in textile industries & savings. https://www.fibre2fashion.com
Ramachandran, T., Karthick, T., & Saravanan, D. (2008). Novel trends in textile wet processing. *IE(I) Journal-TX, 89*.
Rissanen, T. (2005). 'From 15% to 0: Investigating the creation of fashion without the creation of fabric waste' Presenter. *Kreativ Institut for Design og Teknologi*. https://www.ecochicdesignaward.com
Sawhney, A. P. S., Singh, K. V., Sachinvala, N., Li, G., Pang, S. S., & Condon, B. (2006). Successful size-free weaving of cotton yarns on a modern high-speed machine: A progress report. In Beltwide Cotton Production and Research Conferences. Proceedings of the Beltwide Cotton Conferences, National Cotton Council of America (pp. 2491–2496).
Shishlina, N. I., Golikov, V. P., & Orfinskaya O. V. (2000). Bronze age textiles of the Caspian Sea Maritime Steppes, Interpretations of Eurasian archaeology, the bronze age part III. https://www.csen.org/BARBook/BAR.Part01.TofC.html
Velden, N., Patel, M. K., & Vogtländer, J. G. (2014). LCA benchmarking study on textiles made of cotton, polyester, nylon, acryl, or elastane. *International Journal of Life Cycle Assessment, 19*, 331–356.
Vogtländer, J. G. (2013). *Eco-efficient value creation, sustainable design and business strategies*. Delft: VSSD.
Wicker, P., & Becken, S. (2013). Conscientious vs. Ambivalent consumers: Do concerns about energy availability and climate change influence consumer behavior? *Ecological Economics* [e-journal], *88*, 41–48.
Xiong, X., & Li, J. (2010). The degradation of polylactic acid fiber. In *Proceedings of the 2010 International Conference on Information Technology and Scientific Management*. 978-1-935068-40-2 © 2010 scires.

Online Address

http://www.lenzing.com/en/investors/equity-story/global-fiber-market.html
http://www.deutsches-museum.de/en
http://www.etn-net.org/routes/intro/indust_timetable.htm
http://www.gtai.de/gtai/content/en/invest/_shareddocs/downloads/gtai/brochures/industries/industrie4.0-smart-manufacturing-for-the-future-en.pdf
http://bettercotton.org/
http://www.sustainablecotton.org/
http://naturecolored.com/naturally_colored_cotton.htm
http://www.bluecompetence.net/documents/1848986/3365500/rieter%201%20e/dc6d9250-c25b-4ef5-9e5a-d544cb3c590f;jsessionid=1bf3ee7f3eb3e9526fd2fba0b3ac242b
http://www.textileworld.com/textile-world/features/2012/03/spinning-with-an-air-jet/
http://tencelat20.lenzing.com/fileadmin/template/downloads/04_leitner_tencel_at__new_orleans_december_5th_2012.pdf
http://www.dyecoo.com/pdfs/dyecoo-stroy.pdf
http://www.fibre2fashion.com/industry-article/12/1129/energy-conservation-in-textile-industries-savings1.asp
https://www.airdye.com
http://www.enerconind.com/treating/library/applications/plasma-treatment-for-nonwovens-and-textiles.aspx
http://www.elvajet.com/en/applications/fashion/examples/industrial-textile-printing.php
http://www.dyes-pigments.com/natural-dyes.html
http://sewingandstyle.blogspot.com.tr/2012/04/best-software-for-pattern-making.html
http://www.itmf.org/wb/media/conference/bregenz/results/gherzi.pdf

Sustainable Defence Textiles

V.A. Venkatachalam, V.A. Kaliappan and R. Vijayasekar

Abstract Business organisations need to operate sustainably for the global well-being. "Do unto the sustainability as you would have it do unto you" is the "crux", which has to vibrate in every individual's soul of Textiles and Clothing (T&C) stakeholders to ensure that "Textile world meets the needs of the present without compromising the ability of future generations to meet their own needs". This is emphasised through an overview, life cycle, sustainable/unsustainable defence T&C features, sustainable procurement practices, and twenty-first-century realities and best practices. The generic viewpoints discussed are derivable to suite defence T&C.

Keywords Defence · Sustainability · T&C · Life cycle · Procurement · Techniques · ISO · UNEP · Best practices · Standards · Labels

1 Overview on the Sustainability/Unsustainability

Business organisations have relationship to the society and environment in which they operate, and it is a vital factor in their ability to continue to operate sustainably. Sustainability means "to live a life where one is not taking any more from the earth than what one is giving back". "Sustainability is the level at which humans are able to live and co-exist indefinitely with the natural world without harming or causing damage to either side". It is a partnership with nature, a mechanism by which one is reminded to act in sustainable ways. With every scientific and technological accomplishment of today, unsustainability at varied intensities is commingled. Unsustainability phenomenon is not an absolutely unexpected state of affairs. Due to reasons of "knowingly for affluent economic gain or reasons of socioeconomic

V.A. Venkatachalam (✉) · R. Vijayasekar
Textile Technology, Bannari Amman Institute of Technology, Sathyamangalam, India
e-mail: venkatachalama@bitsathy.ac.in

V.A. Kaliappan
PSG College of Technology, Coimbatore, India

S.S. Muthu (ed.), *Textiles and Clothing Sustainability*, Textile Science and Clothing Technology, DOI 10.1007/978-981-10-2474-0_2

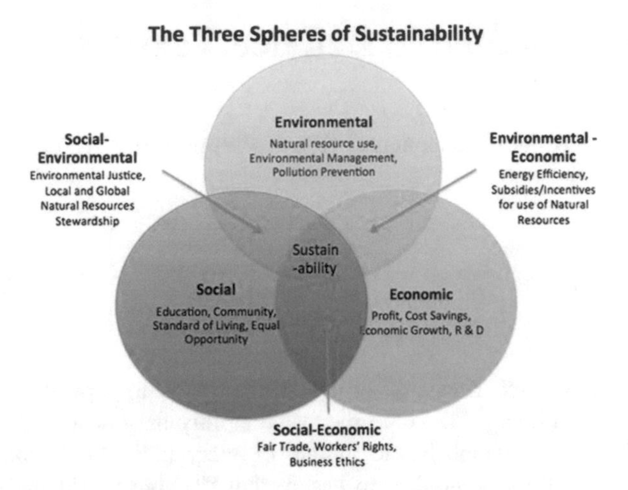

Fig. 1 Inter-relationships of three pillars of sustainability. *Source* https://www.pinterest.com/pin/442478732117449600/?from_navigate=true

coercion or inadvertently by the common people", seeding takes place as a nucleus towards a perilous setting at a later stage. Equanimity of ordinary people is being exploited in isolated situation and which adds to unsustainability at the global level. Sustainability incorporated costing to finalise the selling price is not an unpractical one. Consumers[1] are willing to spend up to 20 % more on environmental sound products and services.

Two thousand years ago, "the adversity what we face today because of environmental unsustainability" was neither known nor predictable; yet the sage Thiruvalluvar has pronounced emphatically in couplets[2] about this distressing environmental problem as stated below:

- Rains fail to fall on the land where people are unjust. When natural resources are plundered, the nature obviously feels it is no use giving such people the natural gifts
- Where there are good people, the environment is well taken care of and the Nature in return takes care of people blessing them with copious rains and plentiful crops

The sustainability is addressed with three pillars, viz. environmental, social and economic.

The interrelationships of these three pillars of sustainability and its upshot are as depicted in Fig. 1.[3]

[1]http://www.eco-officiency.com/benefits_becoming_sustainable_business.html.

[2]http://thirukkuralmadeeasy.blogspot.in/2007/01/environment.html.

[3]https://www.pinterest.com/pin/442478732117449600/?from_navigate=true.

The common principles[4] of sustainable development are as detailed in Exhibit 1:

- Conservation of biodiversity and ecological integrity
- Stable natural capital and sustainable income
- Ensuring intragenerational and intergenerational fair play
- Recognising the global aspects
- Dealing cautiously with risk, uncertainty and irreversibility
- Ensuring appropriate valuation of environmental assets
- Incorporation of environmental and economic objectives in policies and activities
- Social equity and community participation

Exhibit 1: Common principles of sustainable development

Engineering is the driving force of the industrialisation, and this needs to be sustainable to realise the quality of life. The sustainable engineering means[5] "the integration of social, environmental and economic considerations into product, process and energy system design methods". The intent is to minimise environmental impacts across the entire life cycle while simultaneously maximising the benefits to social and economic stakeholders with consideration of the complete product and process life cycle during the design exertion.

Green Economy proposed by "UNEP Green Economy Programme"—means[6]—the production, distribution and consumption of goods and services that result in improved human well-being over the long term, while not exposing future generations to "significant environmental risks and ecological scarcities" through:

- Low-carbon economy
- GDP growth subject to green conditions
- Green-collar jobs
- Circular economy
- Economy subject to ecological principles

In order to create and maintain the sustainability in the society, the four capital models as coined by the University of Melbourne—namely Human Capital, Financial Capital, Environmental Capital and Manufactured Capital, must be

[4]http://www.green-innovations.asn.au/sustblty.htm.

[5]https://www.rit.edu/programs/sustainable-engineering-ms.

[6]https://www.cbd.int/doc/meetings/im/wscbteeb-mena-01/other/wscbteeb-mena-01-unep-green-economy-arab-en.pdf.

balanced. For example, too much attention to human or manufactured capital may affect the environmental sustainability and so on with other capitals.

The declaration in Exhibit 2 is a proverb to be taken into the minds of every stakeholder concerned with the supply chain, in support of not to contribute for unsustainability.

Sustainable Business Model

“+ …aims to do ‘the right thing’ besides ‘doing things right’. (people, planet, profit balanced)”

Jansen, B., van Lieshout, M. (2010) Producten worden diensten: een duurzame waardepropositie voor Vlaanderen!

Exhibit 2: Sustainable business model

Unsustainability elements had been set in motion as of no consequence inevitability in the early industrialisation process (as an aegis for economic development) and in the present millennium, in spite of the UN declaration “2005–2014 as the decade of sustainable development”, it is by now alarmingly on track,[7] towards ecological collapse—the sixth great extinction of life on Earth—and T&C sector is one of the major subscribers to this. Unsustainability is already unleashed by the fact that “exploitation has alarmingly exceeded 1.5 times the restoring gift of nature, i.e. the nature’s capability of refurbishing”. CO_2 level has been increasing around 3 ppm per year, a 20-fold increase compared to pre-industrial times when the highest recorded increase was 0.15 ppm per year; 300–350 ppm of CO_2 is the threshold. These transforms are irreversible on a timescale of human civilisations.

Even if all industrial activity magically ceased today, the footprint has by now left will be felt for eons.[8] “Ecosystem services”[9] that nature performs for free include:

- Soil formation
- Water purification
- Climate regulation
- Pollination
- Nutrient cycling
- Waste treatment, etc.

[7] http://carolynbaker.net/category/collapse-of-industrial-civilization-2/.

[8] https://collapseofindustrialcivilization.com/.

[9] https://www.google.co.in/search?tbo=p&tbm=bks&q=isbn:9290907223.

Total value[10] of ecosystem services, appraised 10 years back as $33.3 trillion dollars, is twofold compared to the then world GDP, a palpable indicator on the benefit from nature every sentient being deriving without any physical and intellectual effort. "Do unto the environment as you would have it do unto you"[11] is a proverb to be recognized sincerely. Should this not be valued and reverenced every day?

The present attitude of many stakeholders is still in line with neoclassical economics as signified in Exhibit 3; each of these can contribute to environmental problems.

- *Resources are infinite or substitutable.*
- *Long-term effects are discounted.*
- *Costs and benefits are internal.*
- *Growth is good.*

Exhibit 3: Neoclassical economics

Minimum compliance as stated below is the prevalent mindset with most of the business communities unless they seriously pursue the concept of CSR, which have an influence on unsustainability:

- Follows the law, but performing not more than required.
- Accomplishing the minimum in order to not to break the law.
- Fulfilling the minimum requirement.
- Following the minimum standard.

To overcome this syndrome, practicing of Hickel's law[12] may dissuade the polluters of any kind who are prone to shatter the sustainability.

Along with the above postulates, the **dynamic concept**[13] of sustainability also needs to concurrently pulsate in the minds of everyone, especially in the views of intellectuals, who are advocating the sustainability's criticality. Societies, environments, technologies, cultures, values, aspirations change and a sustainable society must allow and sustain such change; i.e. it must allow continuous, viable and vigorous development, which is what meant as sustainable development.

[10]https://books.google.co.in/books?id=EuDGQURW6gsC&pg=PA247&lpg=PA247&dq=$33.3+Trillion+dollars&so.

[11]https://books.google.co.in/books?isbn=1895643015.

[12]http://www.counterspill.org/article/santa-barbara-oil-spill-brief-history.

[13]Hartmut Bossel, Indicators for Sustainable Development: Theory, Method, Applications, International institute for sustainable development, 1999.

2 Life Cycle

Life cycle is one of the important components for sustainability/unsustainability discussion.

Life cycle rationalism facilitates sourcing raw materials, manufacture and distribution, use, reuse, recycling, energy recovery and disposal sustainably. As voiced by PACIA,[14] it readily accelerates to:

- Recognise business value
- Identify "hot spots" for investigation
- Develop a map of supply chain
- Collate organisational and supplier information
- Have knowledge on the financial, environmental and social costs of the product over its life cycle
- Fortify relationships and collaboration with mission partners.

Life cycle design approach for resource optimisation[15] as shown in Fig. 2 starts with the fibre stage, looping their way around clockwise and ends with energy recovery.

As discussed by Fletcher,[16] the concepts: (i) negative and (ii) positive feedback loop may be an added input to contain unsustainable textiles from source to final disposal. 80 % plus of textiles environmental impact is knowable at the design stage. Sustainable design, timeless style, effective utilisation of fabric in the production process, choosing apt processing and dyeing techniques along with opportunities for recycling, resource efficiency, and design contest causing lower impact on the environment are certain strategies worth pondering. For example—Levi Strauss's[17] "Repurposed, Reimagined, Resourceful"; a lightweight, yet durable, the parachute trucker jacket-one of the pioneering attempts to reuse disposable nylon military parachutes. "Zero-waste Textile Initiative";[18] and "Turning old into new with reused sustainable fibres"[19] are some more initiatives in the direction of textile sustainability linking to life cycle.

---

[14] http://www.supplychainsustainability.org.au/life_cycle_thinking.

[15] http://www.georgeron.com/2014_05_01_archive.htm.

[16] RS Blackburn, Sustainable textiles—Life cycle and environmental impact, Woodhead publishing limited, 2009.

[17] http://levistrauss.com/unzipped-blog/2014/04/repurposed-reimagined-resourceful-the-parachute-trucker/.

[18] http://sfenvironment.org/news/press-release/san-francisco-launches-zero-waste-textile-initiative-to-keep-apparel-footwear-line.

[19] http://www.triplepundit.com/special/sustainable-fashion-2014/ingenious-reuse-sustainable-fibers/.

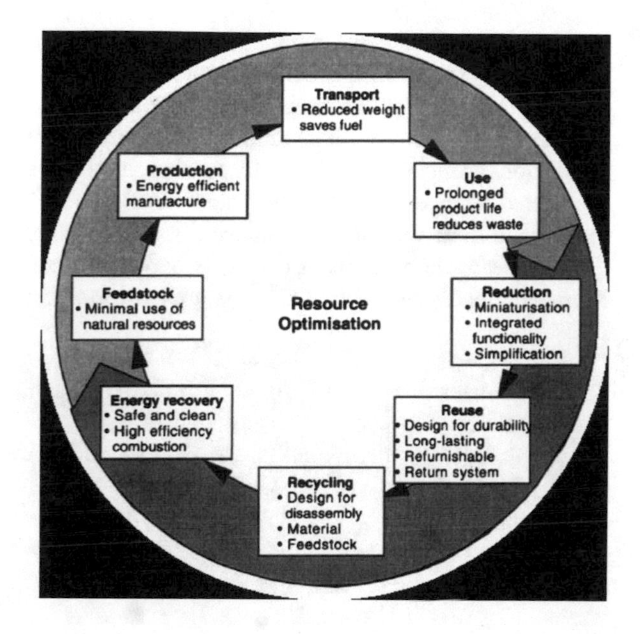

Fig. 2 Resource-optimised cradle-to-cradle life cycle concept. *Source* http://www.georgeron.com/2014_05_01_archive.htm

Unsustainability in the earlier stage of new products is because of partial ignorance due to uncertainties:

- In the conceptual model used for the design calculations,
- In the specific characteristics of the materials purchased,
- Caused by variations in processing,
- About the nature of life cycle the finished product will meet, and
- In various unsustainable unknown factors.

If all the above elements for new product/process are expected to be taken care in the early stage itself, the spirit of new developments will get hampered. These unsustainable elements can be or rather possible to put rightly as signified in Graph 1, mostly just before introduction stage, and the missed out and certain

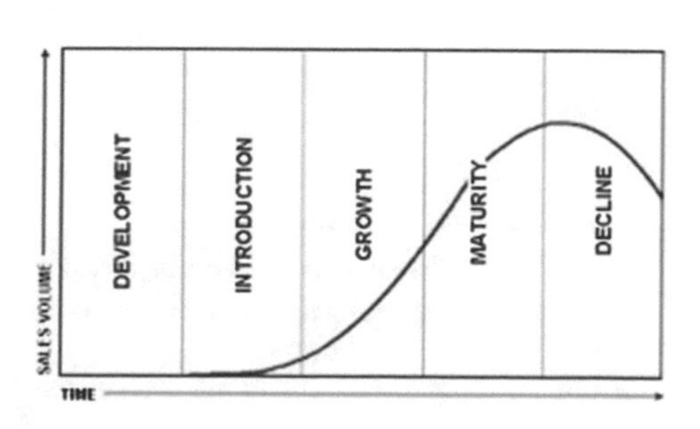

Graph 1 Product life cycle curve. *Source* http://marketingsaffolaactivedairy.blogspot.in/2014_09_01_archive.html

unexpected unsustainable germination during the growth stages, by a systematic inclusive approach. In this context, lawmakers cannot be expected always to keep up with technological development nor would consumers necessarily want to see laws changed with each innovation. Empowerment of concerned associations and inspection agencies will facilitate continuously to ward off the unsustainability associated with every innovation and development through a balanced outlook rather than the minimum compliance on sustainability statute.

The life cycle assessment (LCA) is another important tool for achieving environment friendly products. LCA[20] is a useful tool in:

- Understanding of the important processes within the life cycle
- Identifying weak points and optimisation possibilities of analysed life cycles to further decrease the environmental impacts of the respective products
- Discover measures to effectively reduce environmental impacts
- Preventing the shifting of environmental problems to other stages in the life cycle

Many textile companies have started committing to apply LCA for their manufactured products. LCA[21] is a potentially powerful tool which can give a hand to:

- Regulators to formulate environmental legislation
- Facilitate manufacturers to analyse their processes and improve their products
- Probably enable consumers to make more informed choices

Durability[22] of the textile product drops at each step of the product life cycle due to the impact of processing, service conditions and environmental factors as shown in Graph 2. For this reason, the durability of the military textiles is to be taken care through right approaches at each stage. The efforts underway to enhance the performance of textile materials through nanotechnology and electro-textiles may possibly add a positive value in the defence textile field.

[20] https://en.wikipedia.org/wiki/Life-cycle_assessmentISO 14040[8] and 14044[9] standards.

[21] http://www.tankonyvtar.hu/en/tartalom/tamop425/0032_kornyezetiranyitas_es_minosegbiztositas/ch11s03.html.

[22] http://www.georgeron.com/2014_05_01_archive.htm.

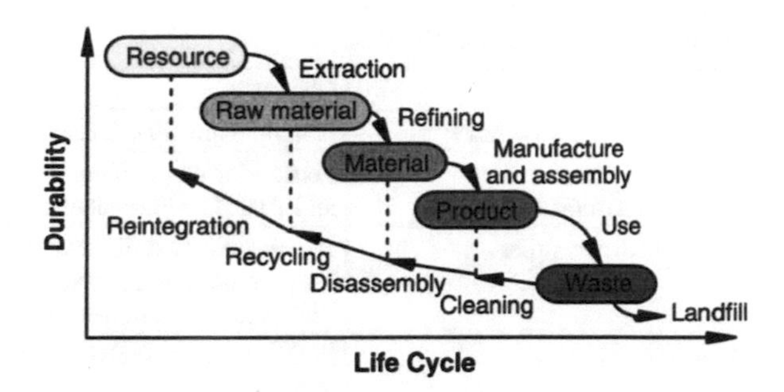

Graph 2 Drop in durability of textiles at each stage. *Source* http://www.georgeron.com/2014_05_01_archive.htm

3 Defence Textile Sustainability

3.1 Sustainable Textile

Sustainable textiles are that meet or exceed the environmental, social and economic performance requirements as set forth by the SMART© Sustainable Textile Standard 2.0,[23] and they have the following attributes:

- All materials and process inputs and outputs are safe for human and ecological health in all phases of the product life cycle
- All energy, material and process inputs come from renewable or recycled sources
- All materials are capable of returning safely to either natural systems or industrial systems
- All stages in the product life cycle actively support the reuse or recycling of these materials at the highest possible level of quality
- All product life cycle stages enhance social well-being

Environmental impact of textiles and garments varies significantly depending upon the type of fibre and related routes. Generally various environmental issues arise in T&C[24] are as stated in Exhibit 4.

- Energy use, greenhouse gas (GHG) emissions, nutrients releases (leading to eutrophication) and eco-toxicity from washing (water heating and detergents) and dyeing of textiles
- Energy use, resource depletion and GHG emissions from synthetic fibre manufacturing, e.g. polyester or nylon

[23]www.sustainableproducts.com/smartweb.html.

[24]Sustainability of textiles ISSUE PAPER No 11, [August 2013], Retail forum for sustainability.

Table 1 Unsustainable chemicals

Purpose	Chemicals
Detergents and auxiliaries	Nonylphenol Ethoxylates (NPEs)
Water, oil, stain and wrinkle-resistant coatings	Perfluorinated compounds (PFCs, including PFOS, PFNA and FTOH)—Formaldehyde
Fire retardant textiles	Poly-brominated diphenyl ethers (PBDEs), hexabromocyclododecane (HBCD)—Short-chain-chlorinated paraffins (SCCPs)—Asbestos
Plastic coatings	Phthalates (e.g. DEHP)—heavy metals (e.g. lead, cadmium and organotins)
Antibacterial and anti-mould agents	Silver—Triclosan—Dimethylfumarate (DMF)
Dyes and colourants	Heavy metals (e.g. mercury, cadmium and lead)—Azo-dyes

- Significant water use, toxicity from fertiliser, pesticide and herbicide use, energy use and GHG emissions associated with fertiliser generation and irrigation systems related to production of fibre crops, e.g. cotton
- Water use, toxicity, hazardous waste and effluent associated with the production stage, including pre-treatment chemicals, dyes and finishes

Exhibit 4: Environmental impact of T&C

The main GHG emissions are CO_2 from energy use, and CH_4 and N_2O from cotton production[25].

The environmental sustainability as thought-out by Multilateral Investment Guarantee Agency[26] talks about waste characteristics, pollution prevention and control, target pollution loads, treatment technologies, emission guidelines, emission levels relating to country specific legislation, monitoring and reporting.

Some of the unsustainable chemicals in use for various textile needs are as given in Table 1.[27]

The opinion by Katie Smith[28] elucidates the challenges and opportunities in T&C as below and this could be expressively accounted for defence textile needs:

---

[25]http://www.sifo.no/files/file78707_oppdragsrapport_1-2013_web.pdf.

[26]https://www.miga.org/documents/Textiles.pdf.

[27]http://www.unep.org/hazardoussubstances/Portals/9/CiP/CiPWorkshop2011/CiP%20textile%20case%20study%20report_21Feb2011.pdf.

[28]http://www.just-style.com/analysis/fact-fiction-and-the-future_id120417.aspx.

- The traditional cotton farming requires large quantities of water, fertilizers and pesticides; the mulesing of sheep and sericulture process is considered unethical and unsustainable.
- The viscose rayon and synthetic fibres, including polyester, polyamide and elastane, have an impact on the environment because of less sustainable resources and production process.
- Technologies such as waterless dyeing, fibre modification and solution or "dope" dyeing, as well as digital printing, have less environmental impact on textile dyeing and printing.
- The eco-friendly dyestuffs reduce water and energy usage.
- Denim finishing techniques such as bleaching and sand blasting are now getting replaced slowly by sustainable lasering and ozone processes.

Buying clothes that have been made from unsustainable means as discussed above is an approval for social and environmental exploitation, leading to unsustainability.

As an example, "Extending the average life of clothes by just three months of active use would lead to a 5–10 % reduction in each of the carbon, water and waste footprints".[29] Making from green route fibres and constructed with durability, reasonable living wages, safer factories, toxic-free, cruelty-free, etc., are considered as sustainable mode. "Greenpeace International" has prompted many T&C units to eliminate toxic chemicals from their supply chains, and chemical companies to introduce greener alternatives, a sustainable progress.

3.2 Military Textile Requirements

Right from the late twentieth century, high-performance man-made fibres well along with natural fibres having been chemically treated/coated/laminated are at task in defence and started planting its share towards unsustainability.

The primary purposes of military clothing always have been protection, functionality, and identification. "Textile learner"[30] has categorised "Features" and "Fibres" for armed forces T&C as in the Exhibits 5 and 6, respectively.

- Lightweight flame retardant clothing with buoyant property for the naval and armed forces personnel to protect them from cold—Nomex fibre
- Head hoods and hand gloves—Kevlar fibre

[29]http://www.ethicalconsumer.org/shoppingethically/ethicalfashion/eco-fashion.aspx.

[30]http://textilelearner.blogspot.in/2014/07/features-and-characteristics-of-armed.html.

- Chemically treated cotton fabrics—considered as the best
- Two ply fabrics—used for the clothing
- Blanket made for the soldiers—two ply fabric made from glass fibre
- Fibrous silica
- Ballistic fabrics for protection from shrapnel
- Flame-resistant clothing for protection from improvised explosive devices (IEDs) and other sources of heat and high energy
- Fabrics for war fighters in the event of chemical or biological attacks—multilayered fabric ensembles allow them to perform critical missions in extreme cold weather
- Moisture wicking fabrics to carry out strenuous tasks for extended periods and with more precision

Exhibit 5: Features of armed forces textiles

- Polyester
- Cotton
- High tenacity polyester
- Lycra
- Kevlar®-Para-aramid
- Coolmax®-specially engineered polyester
- Nomex®-Meta-Aramid

Exhibit 6: Fibres used for making military T&C

Along with the above, some more fibres used for certain specific defence T&C requirements are:

An array of ultra-high modules polyethylene (UHMPE) fibres for Explosive Ordnance Disposal (EOD)—characteristically gel spun polyethylene (GSPE) fibres, with trade names such as Dyneema (DSM) and Spectra aiding reduction in the weight of the garment by about 15 %.

These are also used for cut resistant gloves, helmets and other protective garments. The aramid and Ultra-high-molecular weight polyethylene fibres facilitate excellent body armour requirements. Efforts are on in reducing the weight and bulkiness, and improving moisture management. Besides those, as heat-resistant clothing, needle felt Nomex fibre, ceramic and graphite fibre woven fabrics and silicon rubber-coated fabrics are also used.

In all the wars, T&C have played a vibrant role in providing protection to the soldiers. More than 8000 different items ranging from uniforms and body armour to tents and canteens and battle-dress uniforms (BDUs) are in use for the armed forces.[31]

T&C for military purpose faces a complex set of challenges.[32] They must provide protection, durability and comfort in a wide range of varying climatic conditions and war threats. They need to meet specific protective performance requirements[33] in battlefield tanks, aircrafts, underwater, etc., including the high risk such as gravitational forces during high acceleration–deceleration, extreme temperatures, etc. Defence forces on land, sea and air make use of woven, knitted or non-woven and all the required high-performance functionality, such as "Protection, Comfort and Practicality", have been provided through layers of materials rather than a single garment and that are characterised with:

- Protection against natural and battlefield threats
- Thermo-physiological comfort in extreme weather
- Compatibility of intra- and interclothing components.
- Low weight and bulk materials

Sigrid Tornquist[34] has pointed out that fibre reinforced, insect repellent, modular, ergonomic, chemical resistant, paper thin but super tough, omniphobic T&C is needed for defence operations. "Priorities of various requirements are determined by a number of factors, especially the threats currently being encountered and those that are anticipated in the future".

Some of the countless combat essential T&C items identified by NCTO[35] to support the Armed Forces are as in Exhibit 7.

- Combat and flight uniforms
- Helmets
- Flak jackets
- Gear for extreme weather operations
- Parachutes
- Aircraft fuel cells
- Sandbags
- Tents and shelters

[31]file:///C:/Documents%20and%20Settings/AV/My%20Documents/Downloads/nps62-050312-30%20(3).

[32]Anurag Srivastava, Defence Textiles: Present Scenario and Future Challenges, standards India, vol: 27 no. 5 & 6 12/12 August–September 2013, New Delhi.

[33]http://www.technicaltextile.net/articles/protective-clothing/detail.aspx?article_id=2605.

[34]http://advancedtextilessource.com/2014/11/u-s-army-wish-list/.

[35]http://www.ncto.org/industry-facts-figures/textiles-and-our-military/.

- Sheets
- Blankets and hospital textiles
- Airplane panels
- Ammunition bags/pouches
- Fabric for bulletproof vests/helmets
- Chemical protective suits
- Extreme weather protective fabrics
- Interfacing and lining in apparel and shoes
- Parachutes and parachute harnesses
- Personal flotation devices
- Pontoon bridges
- Rafts
- Ropes and cables
- Ship composites
- Stealth fighter plane graphite fibres
- Wet suites

Exhibit 7: Combat essential items

As revealed by Granch Berhe,[36] wide ranges of woven, coated/laminated fabrics used in defence are as given under:

- Water-repellent, waterproof, wind proof, snow shedding, cold areas clothing
- Sleeping bags with high levels of thermal insulation
- Water vapour permeable for clothing and personal equipment and tents
- Rot resistant for tents, covers, nets
- UV resistant to strong sunlight
- Air permeable to hot tropical climates
- Biodegradable discarded and buried fabrics
- Nuclear, biological and chemical protection through activated carbon
- Snow camouflage, fire resistant, moisture management, anti-odour/anti-microbial, taken care through specific fibres and chemicals as basic constituents
- Using fabrics of lightweight, low bulk, high durability and dimensional stability to manage with minimal space available, reliably in adverse conditions for long periods of time without maintenance as the specific defence needs

[36]http://www.slideshare.net/GranchBerheTseghai/7-military-textiles-52380790™.

- Multi-layer fabrics to serve diverse inevitabilities
- Combat fabrics made from non-thermoplastic fibres to minimise melt/burn injuries (extremely important)

The specific properties chartered by Arunabh Chowdhury[37] for defence applications are as detailed in Exhibit 8

- Tensile properties.
- Resistance to water and saline water.
- Resistance to chemicals.
- Camouflaging effects.
- Resistance to fire and high temperature.
- Resistance to infrared detection.
- Resistance to UV and other electromagnetic radiation.
- Resistance to ballistic impacts.
- Resistance to microbiological growth and degradation.

Exhibit 8: The specific properties for defence T&C

Requirements for advanced integrated combat clothing system[38] classified as physical, environmental and physiological are as in Exhibit 9.

Physical

- Durability to prolonged exposure to inclement weather and heavy wear
- Good tensile and tear strength and abrasion resistance

Environmental

- Water repellency, wind proof
- Battlefield (good camouflage and low-noise generation)

Physiological

- Low weight, easy to wear, minimum heat stress
- Air, moisture and vapour permeability
- Comfort and good appearance

[37]Arunabh Chowdhury, Technological Applications of Textile in Defence and Standardization Status, standards India, vol: 27 no. 5 & 6 12/12 August−September 2013 New Delhi.

[38]http://www.nitracoeprotech.org/pdf/status-report-on-protective-textiles.pdf.

Exhibit 9: Requirements for advanced integrated combat clothing system

Every single military T&C necessitates meticulous specifications and for example the parachute[39] fabric for air force to be characterised with:

- Elongation,
- Elastic recovery,
- Energy absorption,
- Porosity,
- Air permeability,
- Strength properties,
- Temperature properties,
- Ageing properties, and
- Impact loading.

Integration of some of the above functionalities through fewer layers to provide multi-layer protection crowns into reduction in life cycle costs because of few components and also effective, durable and recyclable nature. Blends such as poly-cotton and poly-viscose are sustainable, due to their specific qualities such as reduced laundering costs and enhanced fabric durability. Nanotechnology-based fabrics for camouflaging and muscles stimulation are in progress, which may also leverage the sustainability philosophy. Defence textiles characterised with certain specific features met by a range of fibres, chemicals, and technical and management processes, which for sustainability inevitabilities to be suitably regarded encompassing the pillars of sustainability and duly accounting whole life cycle. The issues, barrier to sustainability requirements, need to be addressed strategically by the defence T&C authorities. The clothing of the service personnel as such assumes greater significance when uniformity and aesthetics is looked upon, besides functional aspects. Nevertheless, while the socio-economic factors with aesthetic desire aspects influence the civilian requirements, the defence T&C is inclined more on functional and technical quality rather than on stringent cost factor. Because of this reason, military requirements such as uniforms, callisthenics clothing, extreme climate garments, back packs, sleeping bags, tents equipment covers, bags and more such products are at an advantageous status and hence could be handled in a sustainable supply chain route with less of economic constraints.

Clothing and accessories manufactured from natural and fair trade materials like soy, organic cotton, bamboo, and leather alternatives are becoming eco conscious, socially responsible, stylish and at the same time kind to our planet. This may be a meaningful choice relating to specific demanding military requirements. Most protective clothing made of inherently and permanently non-flammable aramid fibres with their ability to be recycled is viewed as sustainable. The whole product's

[39]http://lhldigital.lindahall.org/cdm/ref/collection/parachute/id/1379.

life cycle should be regarded in an integrated perspective; the defence T&C representatives from design, development, production, marketing and utilisation should work together on the eco-design of new product. As a group, they have the best chance to realise the holistic effects of the product on environmental impact.

According to the National Society of Professional Engineers, the most important ethical principle is to "hold paramount the safety, health, and welfare of the public (society)", which may be aptly extended for other sustainability pillars too. Defence textile research may adopt innovative inquest by recounting the saying by Danish Environmental Protection Agency (EPA), "for all uses and in all circumstances a suitable less toxic alternative can be found".

4 Sustainable Procurement

4.1 Sustainable Procurement—What It Means?

Sustainable procurement[40] "is not about inconveniencing the market with extra requirements; rather, it is a well-defined strategy that gradually gets into sustainable requirements in bids, supports measures and promotes dialogue and open communication between the suppliers and procurers".

The techniques suggested by Rio2016[41] for a typical context as detailed below may be mapped to match the defence-specific T&C deal.

- Substituting sulphuric acid with CO_2
- Digital textile printing
- Collection of clothing manufacturing leftovers and send to shredding machines
- Recovery of re-evaporation latent heat
- Captured heat used for drying sludge generated
- A reverse logistics programme for proper disposal of unserviceable old textiles
- Development of natural finish
- Carbon footprint optimisation
- Precautionary approach to chemicals management
- Development of supply chain policy
- Evaluation and monitoring of the supply chain

[40]https://www.ungm.org/Areas/Public/Downloads/Env_Labels_Guide.pdf.

[41]http://www.rio2016.com/sustentabilidade/wp-content/uploads/2016/02/Rio-2016-Sustainable-Textile-Guide.pdf.

ISO 20400[42]—The new international standard for sustainable procurement practice likely to bring value beyond the procurement by helping to disseminate CSR practices contained in ISO 26000:2010 Guidance on social responsibility, throughout supply chains, and ultimately the entire economy. The structure of the standard in line with BS 8903 is as follows:

- **Fundamentals**—the key drivers and principles of sustainable procurement
- **Policy and strategy**—key issues to consider in developing a policy and strategy
- **Organising the procurement function**—creating the organisational conditions necessary to procure sustainably and setting priorities
- **Procurement process**—alternative modus operandi for sustainable procurement

BS 8903[43] provides guidance to any size and type of organisation on adopting and embedding sustainable procurement principles and practices. It covers all stages of the procurement process which may be of interest to defence T&C.

Sustainable public procurement (SPP) includes the three pillars of sustainable development embracing transparency, fairness and non-discrimination: competition, accountability and verifiability. The SPP benefits[44] are as revealed in the Exhibit 10.

- Climate change and or greenhouse gas emission reduction;
- Ozone-depleting substances eradication
- Natural resource use optimisation
- Waste minimisation
- Job creation
- Equality and diversity
- Fair pay for supply chain workforce
- Economic regeneration
- Legal compliance
- Public image enhancement

Exhibit 10: Sustainable procurement

[42]http://www.iso.org/iso/home/news_index/news_archive/news.htm?refid=Ref1873.

[43]http://actionsustainability.itineris.co.uk/evaluate/procurement-standard.aspx.

[44]http://www.unep.org/resourceefficiency/Portals/24147/scp/procurement/docsres/ProjectInfo/UNEPImplementationGuidelines.pdf.

4.2 *Sustainable Procurement Factors*

Sustainable procurement is influenced as listed underneath:

(i) Policies and processes
(ii) Approaches
(iii) Key concepts
(iv) Barrier for SPP
(v) Strategies
(vi) Life cycle thinking
(vii) SPP tools
(viii) SCP indicators
(ix) Feedback loop

(i) Policies and Processes[45]

Sustainable procurement,[46] emerged from Rio Earth Summit in 1992, can be allied to defence with an outlook to maximise sustainable benefits for themselves and the whole world. This goes beyond the upfront cost, encompassing associated costs, environmental and social risks, benefits and implications. It means[47] "defence procurers meet their needs for goods, services, works and utilities in a way that achieves value for money on a whole life cycle basis in terms of generating benefits not only to them, but also to society and the economy, while minimising damage to the environment".

(ii) Approaches

There are **two approaches** as detailed below, which may be appositely considered by the defence procurers:

(i) Product-based: This is commonly used when an organisation wishes to understand the impact of a product or product range and assesses the environmental credentials for strategic and marketing purposes.
(ii) Supplier-based: Implemented effectively, this method will show whether a supplier meets the environmental standards of the organisation, along with meeting the requirements of law.

---

[45]http://www.gao.gov/key_issues/leading_practices_acquisition_management/issue_summary.
[46]https://en.wikipedia.org/wiki/.
[47]http://www.unep.org/resourceefficiency/Consumption/SustainableProcurement/tabid/55550/Default.aspx.

(iii) Key Concepts

- Value for money
- Environmental buying
- Supply chain assessment
- Life cycle thinking

(iv) Barrier for SPP/GPP

Economy, policy, procurement and market-related barrier for SPP/GPP in the ascending importance ascertained by UNEP[48] through a survey are as follows:

- *Difficulty in Audit office approval*
- *Protecting budding industries*
- *Concern about quality of sustainable products*
- *Lack of suppliers of SPP/GPP products*
- *Inadequate public procurement system for incorporating SPP/GPP*
- *Insufficient supply of sustainable goods and services*
- *The initial cost commitment*
- *Lack of clarity on sustainable product*
- *Budgetary/resource restrictions*
- *Lack of legislation or regulations*
- *Lack of interest and commitment from procurers*
- *Lack of technical capacities on environmental/social issues*
- *Lack of information and knowledge about SPP/GPP*
- *Perception that sustainable products are more expensive*

These barriers need forestalling through a proper system approach before realising the value of SPP.

(v) Strategies

As guided by IISD,[49] the following **strategies** may be suitably adopted for sustainable military T&C:

- By-product synergy and industrial ecology
- Cleaner production

---

[48]http://www.unep.org/resourceefficiency/Portals/24147/SPP_Full_Report_Dec2013_v2%20NEW%20(2).pdf.

[49]https://www.iisd.org/business/tools/bt.aspx.

- Design for environment
- Eco-efficiency
- Energy efficiency
- Environmentally conscious manufacturing
- Reduction, reuse, recycling and recovery
- Green procurement
- Performance contracting
- Pollution prevention
- Zero-emission processes

Clear, transparent and consistent policies and processes are the crux to implement strategic decisions. Engaging with suppliers who have commitment to best practice, continuous improvement and collaborative multi-supplier approaches ensure sustainability performance with a partnership.

(vi) Life Cycle Thinking[50]

As UNEP explains "it is about going beyond the traditional focus on production sites and manufacturing processes, i.e. taking into account the environmental, social and economic impact of a product over its entire life cycle, including the consumption and end of use phase". Whatever is the business size, life cycle thinking in supply chain sustainability leads to:

- Ascertain business value
- Identify "hot spots" for further investigation
- Develop the map of supply chain to one's needs
- Collate organisational and supplier information
- Understand the financial, environmental and social costs of textile and clothing over their life cycle
- Strengthen relationships and collaboration with supply chain partners
- Achieve greatest return.

Life cycle thinking facilitates all jointly being proficient from sourcing raw materials to manufacture and distribution, and from use or consumption to reuse, recycling, energy recovery and disposal.

[50]http://www.supplychainsustainability.org.au/life_cycle_thinking.

(vii) SPP Tools

A range of **tools**[51] practiced by PACIA, as given below, may be fittingly considered by the defence establishment in their T&C procurements to realise financial, environmental and social benefits:

- Life cycle map,
- Assessment matrices,
- Carbon foot printing,
- Energy mass balance,
- Life cycle costing,
- Environmental life cycle assessment, and
- Social life cycle assessment.

(viii) SCP Indicators

The **SCP indicators**[52] identified beneath are additional means to back SSP:

- Land use and biodiversity
- Socio-economic
- Waste and pollution
- Material consumption and resource use

(ix) Feedback Loop[53]

For sustainability purpose, corporations need to expand their information-monitoring systems to actively collect a broad range of ecological feedbacks in addition to social feedback.

4.3 Sustainability Reporting Initiatives[54]

A sustainability report of an organisation unfolds the economic, environmental and social impacts caused by its activities along with the organisation's values and governance model, and its commitment to a sustainable global economy. The

[51]http://www.pacia.org.au/docs_mgr/PACIA__8StepGuide_to_SupplychainSustainability.pdf.

[52]http://www.unep.org/pdf/SCP_Poverty_full_final.pdf.

[53]http://www.ecologyandsociety.org/vol2/iss2/art12/.

[54]https://www.globalreporting.org/information/sustainability-reporting/Pages/default.aspx.

sustainability reporting enables organisations to enhance their sustainable operations. The defence T&C operations may look for this initiative to ensure their sustainability requirements. The various initiatives are as elucidated in infra 9.5.2.

4.4 Sustainable Public Procurement Codes

Few codes as elucidated underneath play a positive role for SSP.

- The Sustainable Supply Chain Management SSCM[55] Code of Conduct supports positively the SPP. Sustainable supply chain management[56] and sustainability strategy[57] (embracing societal change, environmental solutions and better financial futures) and action priority[58] as recommended by Westpac is a green pasture for sustainable procurement.
- The United Nations Global Compact (UNGC),[59] an international corporate sustainability initiative, encourages the growth of responsible businesses through the adoption of ten high-level sustainability principles addressing human rights, labour standards, environment and anti-corruption.
- Green Public Procurement (GPP),[60]—Guides on environmental considerations at each stage of the procurement process towards sustainable practice from product design to disposal, which can be useful for defence T&C related stakeholders. GPP criteria aim to attain a good balance between environmental performance, cost considerations, market availability and ease of verification; defence authorities may choose, according to their needs and ambition level, to include all or only certain requirements in their tender documents.
- SMART's "Sustainable textile standard" is to help raw material suppliers, converters, manufacturers and end users, and to address the triple bottom line, throughout the supply chain. "Sustainable Textile Supply Chain Achievement Matrix"[61] guides to attain a systematic progress towards sustainability.

The above tools and advices may possibly prop up to achieve sustainability in defence T&C procurement.

---

[55]http://radiorusak.esy.es/westpac-s-sustainable-supply-chain-management-code-of-conduct.pdf.

[56]https://www.westpac.com.au/about-westpac/sustainability-and-community/sustainability-action/suppliers/sustainable-supply-chain/.

[57]https://www.westpac.com.au/docs/pdf/aw/SSCM_Framework.pdf.

[58]https://www.westpac.com.au/docs/pdf/aw/SSCM_CodeofConduct.pdf.

[59]https://en.wikipedia.org/wiki/United_Nations_Global_Compact.

[60]http://ec.europa.eu/environment/gpp/eu_gpp_criteria_en.htm.

[61]SMART© Sustainable Textile Standard 2.0—sts-v2.

4.5 Sustainability Certifications

The following sustainability certifications may be considered as reinforcement in the defence T&C procurement process:

- **Global Reporting Initiative (GRI)**[62]—An international-independent organisation that helps businesses, governments and other organisations understands and communicates the impact of business on critical sustainability issues such as climate change, human rights, corruption and many others.
- **Higg Index**,[63] from Sustainable Apparel Coalition (SAC)—A sustainability measurement tool for apparel companies to measure the impacts of their products across the value chain—single approach for measuring sustainability for all buyers without investing in multiple sustainability technologies and certification processes. It is a holistic self-assessment tool empowers brands, retailers and facilities of all sizes, in their sustainability journey, to measure their environmental and social and labour impacts and identify areas for improvement.
- **OEKO-TEX**[64]—focuses on harmful chemicals; specific certifications include "OEKOTEX 100 Standard" and Sustainable Textile Production (STeP)[65]—A certification system to recognise brands, retail companies and manufacturers within the textile chain to communicate their path to sustainable production to the public in a transparent and credible manner.
- **Global Organic Textile Standard (GOTS)**[66] A worldwide standard for processing organic fibres. It uses ecological and social criteria that cover the entire textile supply chain.
- **Bluesign**[67]—brings to light resource efficiency, emissions to air and water, and consumer health and safety; applicable to brands and manufacturers, as well as chemical suppliers—but earmarked towards larger industry players.
- **Better Cotton Initiative (BCI)**[68]—promotes and maintains a set of holistic standards for the cotton supply chain, covering environmental, social and economic sustainability.
- **Cradle-to-Cradle (C2C) Certified**[69]—guides product designers and manufacturers towards creating products that use safe and reusable

[62]https://www.globalreporting.org/resourcelibrary/2012-2013-Sustainability-Report.pdf.

[63]http://apparelcoalition.org/the-higg-index/.

[64]https://www.oekoNtex.com/en/manufacturers/manufacturers.xhtml.

[65]http://www.testex.com/en/downloads/STeP_by_Oeko-Tex/en/STeP-FAQ_en.pdf.

[66]http://www.global-standard.org/the-standard.html.

[67]http://www.bluesign.com/industry.

[68]http://bettercotton.org/aboutNbetterNcotton/betterNcottonNstandardNsystem/.

[69]http://www.c2ccertified.org.

materials, use clean and renewable energy, protect water supplies, and advance social and environmental justice.

5 T&C Sustainability in Twenty-First Century

In conceptual age, i.e. in the twenty-first century the sustainability is at perilous phase because of continued unmindful industrial activity-lacking concern for sustainable pillars. Inter alia, the following techniques, in which some are in vogue in other engineering fields for their sustainability, may be tried in support of sustainable twenty-first-century defence T&C requirements:

1. Clean by design
2. Global reporting initiative
3. Standards and labels
4. Eco-labels
5. Quality function deployment (QFD)
6. Operational research (OR)
7. Cost-benefit analysis
8. Lean manufacturing

5.1 Clean by Design[70]

As said by NRDC,[71] manufacturing practices in some countries and in a number of industries are generally less efficient, using more energy, water and materials than

[70]http://www.ecouterre.com/nrdcs-clean-by-design-helps-chinas-garment-industry-turn-a-corner/.
[71]https://www.nrdc.org/resources/clean-design-apparel-manufacturing-and-pollution.

necessary. Even marginal improvements in manufacturing are expected to deliver admirable cost savings in energy, water and chemical usage, leading to improved sustainability. It motivates to improve process efficiency and to reduce waste and emissions aiding for sustainable environment. Informed decisions as recognised from ethical theory in four impact areas, namely Raw Materials, Manufacturing, Transportation and Consumer Care, facilitate to reduce unsustainable footprint. The eight potential starting points as in Exhibit 11 are capable of increasing profitability at the same time advances sustainable textile manufacturing.

(i) Undertaking failure analysis when things go wrong
(ii) Standardising optimal methods and recipes
(iii) Substituting enzymes in pre-treatments
(iv) Investigating opportunities to reduce salt in dyeing recipes
(v) Increasing reliance on higher fixation dyes
(vi) Improving machine utilisation
(vii) Scheduling optimally to minimise cleaning in between each batch of processing
(viii) Monitoring continuously to check whether implementation of improvements is in place

Exhibit 11: Promising starting points

Cleaner Production (CP)[72] as suggested by Rio 2016 facilitates the realisation of:

- Waste elimination
- Minimisation or elimination of raw materials and other inputs impacting the environment
- Increased energy efficiency
- Reduction or elimination of waste and emissions
- Reduction of pollution
- Reduction in costs of waste management
- Minimisation in environmental liabilities
- Increased health and safety
- Development in environmentally friendly products
- Enhancement in company's image
- Increase in productivity

[72]http://www.rio2016.com/sustentabilidade/wp-content/uploads/2016/02/Rio-2016-Sustainable-Textile-Guide.pdf.

Techniques professed by Oregon, Department of Environmental Quality,[73] also avoid, eliminate or reduce the creation of pollutants at the source. The benefits to business are:

- Reduced operating costs
- Reduced compliance costs
- Reduced liability
- Increased productivity
- Increased marketability as a "green" business
- Even possibly increased profits

5.2 *Sustainability Reporting Initiatives*

Major sustainability reporting guidance[74] includes:

(i) Global Reporting Initiatives (GRI's Sustainability Reporting Standards)
(ii) The Organisation for Economic Co-operation and Development (OECD guidelines for Multinational Enterprises)
(iii) The United Nations Global Compact(UNGP)
(iv) ISO 26000, International Standard for social responsibility

(i) Global Reporting Initiative[75]

A sustainability report measuring, disclosing and accountable to internal and external stakeholders aimed at organisational performance in the direction of sustainable development. The GRI Guidelines are intended to be applicable to organisations of all sizes and types operating in any sector. However, they were developed primarily with the needs of larger businesses in mind.

[73]http://www.deq.state.or.us/programs/sustainability/10ways-businesses.htm.
[74]https://www.globalreporting.org/Information/about-gri/Pages/default.aspx.
[75]Sustainability Reporting Guidelines, © 2000–2006 Global Reporting Initiative, Version 3.0.

(ii) OECD[76]

OECD Guidelines for Multinational Enterprises are far-reaching recommendations for responsible business conduct. It says, "Inspiring the suppliers may be a step forward in the direction of sustainability" and the practices are as indicated under:

- Mapping the impacts and set priorities, and selecting useful performance indicators
- Measuring the inputs used in production, assessing operations of the facility and evaluating products
- Understanding measured results and improving performance

(iii) The United Nations Global Compact[77]

The United Nations Global Compact is a United Nations initiative to encourage businesses worldwide to adopt sustainable and socially responsible policies, and to report on their implementation. It is a principle-based framework for businesses, stating ten principles in the areas of human rights, labour, the environment and anti-corruption.

(iv) ISO 26000[78]

ISO 26000 provides guidance on how businesses and organisations can operate in a socially responsible way. This means acting in an ethical and transparent way that contributes to the health and welfare of society.

(a) Eco-functional Index[79]

Eco-functional Index proposed by Subramanian Senthilkannan Muthu et al., encompassing functional and ecological properties combined with consumer behaviour and covering inputs (raw materials, process of manufacture, functional properties and ecological properties) and outputs (quality, functionality, 3R s, human impact and environmental impact), would be serving the interest of designers, manufacturers and downstream players.

---

[76]http://www.oecd.org/innovation/green/toolkit/actionstepsforsustainablemanufacturing.htm.

[77]https://en.wikipedia.org/wiki/United_Nations_Global_Compact.

[78]http://www.iso.org/iso/home/standards/iso26000.htm.

[79]http://www.sciencedirect.com/science/article/pii/S1470160X1200369X.

5.3 Standards and Labels

5.3.1 Standards

ISO 14000 family[80] of standards comprising the following categories acts as practical tools for to manage sustainable responsibilities:

- Environmental Management Systems (ISO 14001)
- Environmental Auditing (ISO 14010, 14011, 14012)
- Environmental Labelling (ISO 14024)
- Environmental Performance Evaluation (ISO 14031)
- Life Cycle Assessment (ISO 14040)

ISO 14000[81] standards ensuring sustainable development and addressing triple bottom line comprise:

- *Environmental management systems (EMS),*
- *Auditing standard, environmental performance,*
- *Environmental labels and declarations,*
- *Life cycle assessment (LCA),*
- *Greenhouse gas (GHG) accounting and verification,*
- *Environmental communication,*
- *Environmental aspects in product standards,*
- *Eco-efficiency assessment, Material flow cost accounting (MFCA),*
- *Carbon footprint of products,*
- *Eco-design, and*
- *Quantitative environmental information.*

These tools lead to:

- *Reduced raw material/resource use,*
- *Reduced energy consumption,*
- *Improved process efficiency,*
- *Reduced waste generation and disposal costs, and*
- *Utilisation of recoverable resources.*

[80]http://www.iso.org/iso/iso14000.

[81]www.iso.org/managementstandards.

ISO 14001, the Environmental management system (EMS) requires companies to commit to pollution prevention and continual improvement as part of the usual cycle of business management, which is an en route for sustainability.

ISO 14044:2006, Environmental management—life cycle assessment—requirements and guidelines is designed for the preparation, conduct and critical review of life cycle inventory analysis.

The main benefits of implementing an EMS and subsequent certification of the system are:

- Improved internal organisation with clear responsibilities,
- Compliance with environmental expectations of customers,
- Access to new markets,
- Good relationship with the community,
- Waste minimisation,
- Materials and energy conservation,
- Process improvement/increase in productivity,
- Better environmental performance.

ISO 26000:2010[82]—Guidance on social responsibility, which is already being followed by some MNCs,[83] contributes to sustainable development. It encourages industries to go beyond legal compliance and an essential part of social responsibility. It promotes common understanding in the field of social responsibility and complements other instruments and initiatives. This along with "United Nations Global Compact's" ten (see footnote 83) universally accepted principles such as human rights, labour, environment, anti-corruption initiative are excellent guidelines to realise sustainability requirements for defence T&C.

The ISO 14000 family, which comprises a number of standards, complements ISO 14001-2015 and with ISO 9001, SA 8000, OHSAS 18001, REACH, etc., may be considered as added twenty-first-century tools for sustainable T&C management.

ISO 14001:2015 the environmental management system—T&C organisations get benefitted. ISO 9001 Certification provides the foundation for better customer satisfaction, staff motivation and continual improvement. 2015 version of ISO 9001[84] Standard focuses on documentation of implementation of processes and as evidence that activities were performed. It includes three basic core concepts:

[82]http://www.iso.org/iso/catalogue_detail?csnumber=42546.

[83]http://www.iso.org/iso/home/news_index/news_archive/news.htm?refid=Ref1997.

[84]http://www.ursindia.com/iso_9001/hosur/textile%20and%20textile%20products.aspx.

(a) *Process approach*
(b) *Plan-do-check act—methodology*
(c) *Risk-based thinking*

SA8000 is an auditable certification standard that encourages organisations to develop, maintain and apply socially acceptable practices in the workplace. OHSAS 18001[85] is an Occupation Health and Safety Assessment Series for health and safety management systems. It is intended to help an organisation to control occupational health and safety risks. REACH[86] shifts the onus from public authorities to industry with regard to assessing and managing the risks posed by chemicals and providing appropriate safety information for their users.

5.4 ISO Environmental Labels

Environmental labels are aimed at progressive sustainable patterns of production and consumption. The textile sustainability labels help to make the production sustainable coupling with increase in efficiency and profitability of the manufacturers. There are companies that have already started marking green products and include some environmental indicators such as life cycle assessment, water footprint, CO_2 emissions pertaining to their products manufactured, in their website or product labelling. They inform their consumers about their manufacturing process.[87]

The ISO has categorised[88] environmental labels as Type I, Type II and Type III.

Type I: Awards license that authorises the use of environmental labels on products indicating overall environmental advantages of a product within a particular product category, based on life cycle considerations.
Type II: Informative environmental self-declaration claims—refers to an environmental aspect of a product, to a component of the product or to its packaging.
Type III: Voluntary programs that provide quantified environmental data for a product under pre-set categories of parameters are defined by a qualified third party, based on a life cycle assessment, and verified by that or another qualified third party.

[85] http://www.ohsas-18001-occupational-health-and-safety.com/what.htm.
[86] http://ec.europa.eu/growth/sectors/chemicals/reach/.
[87] https://www.diva-portal.org/smash/get/diva2:704976/FULLTEXT01.pdf.
[88] http://www.sustainabilitydictionary.com/eco-labels/.

5.5 *Eco-labels*

Eco-labels—a subgroup[89] of environmental labels focuses on all aspects of the "life" of a product, from design, production, operation and maintenance to disposal. An eco-label is "a visual communication tool indicating environmentally preferable products, services or companies that are based on standards or criteria".[90] And of the 309 eco-labels identified worldwide, 41 cover textiles.[91] UNEP has categorised[92] the eco-labels with implementation Guidelines.

5.6 *Sustainability and Quality Function Deployment (QFD)*

Quality function deployment (QFD)[93] is a widely used TQM tool for translating the Voice of Customer (VOC) into such as technical and sustainability components. The main bottom line of the QFD approach is how to construct the house of quality (HoQ) including how to identify the correlation between the customer needs including sustainability aspects (normally called Whats) and the engineering characteristics (normally called Hows). The development of HoQ would involve cross-functional team or members from different departments in industry.

A typical generic HoQ is depicted as in Fig. 3, in which, well along with other criteria, T&C sustainability-specific factors can be posted to get the optimum benefit.

The relationships and correlations are as portrayed in Fig. 4 and this exercise is capable of addressing the sustainability needs related to development of T&C, utilisation of resources and up to sustainable disposal. Figure 5 is an illustration of HoQ profile to incorporate sustainability requirements.

5.7 *Operational Research (OR) and Sustainability*[94]

Operational research addresses problems in the sustainability domain such as energy and water usage efficiency, scarce resource management, reverse, recycling,

[89]https://www.ungm.org/Areas/Public/Downloads/Env_Labels_Guide.pdf.

[90]http://www.keystone.org/spp/environment/Green-Products-Roundtable.

[91]http://www.fibre2fashion.com/industry-article/5388/creating-a-global-vision-for-sustainable-textiles?page=1#.

[92]http://www.unep.org/10yfp/Portals/50150/10YFP%20SPP/UNEPImplementationGuidelines.pdf.

[93]http://www.ijmmm.org/papers/017-E108.pdf.

[94]http://www.journals.elsevier.com/european-journal-of-operational-research/call-for-papers/call-for-papers-trends-in-operation.

The House of Quality's Structure

The HoQ consists of multiple "rooms":

- **What's**,
  - the *voice of the customer* (i.e., internal, intermediate and ultimate customers)
  - other from regulatory standards
- **How's**, how we are going to satisfy the What's
- **Why's**, customer's perception relative to competition
- **How Much's**, objective targets
  - for assuring requirements have been met
  - inputs for downstream phases
- and the correlations between them.

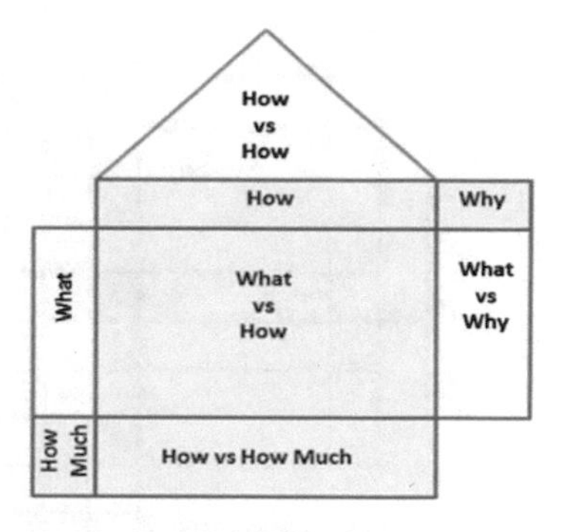

Nicola Mezzetti, Ph.D. Human-Centered Design

Fig. 3 Generic HoQ. *Source* http://www.ijmmm.org/papers/017−E108.pdf

The House of Quality: Relations

Different types of relations:

- **Relationships**: degree of interdependence between What's and How's
  - ⊙ **Strong** relationship
  - ○ **Medium** relationship
  - △ **Weak** relationship
- **Correlations**: describe which of the HOWs support one another and which are in conflict.
  - ⊕ **Strong** positive
  - + **Medium** positive
  - − **Medium** negative
  - ⊖ **Strong** negative

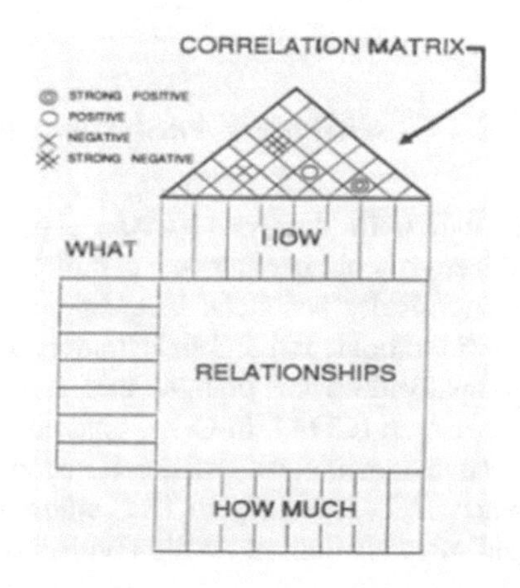

Nicola Mezzetti, Ph.D. Human-Centered Design

Fig. 4 HoQ relationships and correlations. *Source* http://www.slideshare.net/nicolamezzetti/humancentered-design-customer-focus-requirements-engineering

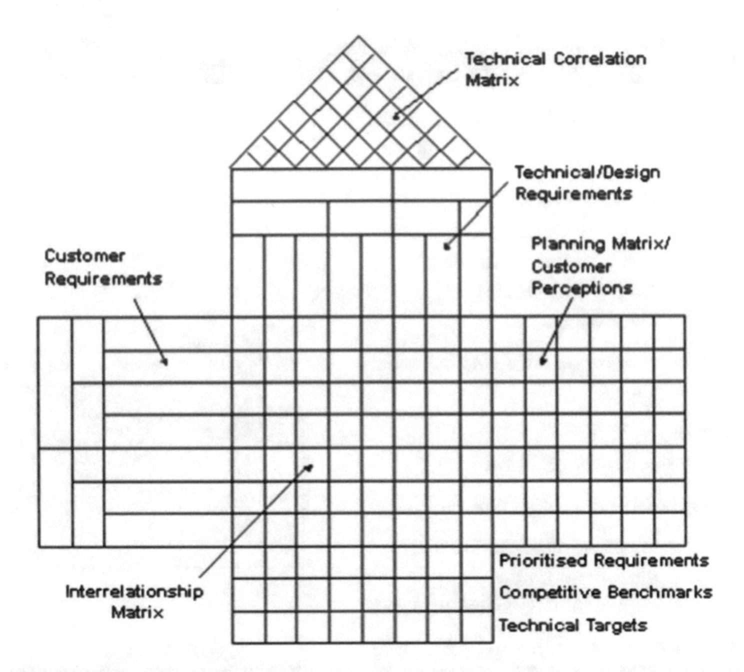

Fig. 5 HoQ profile to place sustainability requirements. *Source* https://blog.cognizantzdlc.com/category/automated-lifecycle-engineering/page/2/

remanufacturing and waste management, reducing carbon emissions, disaster management and emerging problems such as migration crisis management. For scheduling, vehicle routing, facility location, supply chain design, warehouse management, capacity expansion and production planning problems, OR will show the way to an effective T&C-specific sustainable management.

5.8 Cost-Benefit Analysis and Sustainability

Cost-Benefit Analysis (CBA) Approach: In this utilitarianism approach, the objective is not to achieve a completely clean environment, but rather to achieve an economically beneficial level of pollution with human health and environmental considerations. If T&C stakeholders do not take this in their mind and if individual company does not practice this, it will only force to resort into **Cost-Oblivious Approach (COA)**. In COA, cost not taken into account; only the acceptable level of environmental degradation is criterion—such as in the rights and duty ethics. Not practicable, especially in T&C where resources are not infinite and that will be the end of industrialised textile civilisation.

5.9 *Lean*[95] *and Sustainability*

Lean manufacturing is derived principally from the Toyota Production System (TPS). Lean emphasises efficiency, reducing cost and time, and action. It benefits operational, administrative and strategic improvements. It is a systematic method for the elimination of waste created through "overburden" and "unevenness in workloads" (enemies of sustainability) within a manufacturing system.

Waste as shown in Fig. 6 principally does not need to exist. Lean helps the environment without really intending to—in a sense, more efficient production means, less energy, less material, less pollution to air and water, and less hazardous solid waste—which leads to eliminating pollution at the source rather than costly "end of pipe" controls. As waste is eliminated, quality improves while production time and cost are reduced, aiding to sustainability. It is a matter of rethinking to find new uses for the waste material.

Lean is a set of "tools"[96] that give support for identification and steady elimination of waste. The number and type of tool depends on the type of product and production system. A non-exhaustive list of lean tools would include:

SMED—a rapid and efficient way for converting a manufacturing process from the current product to running the next product.

Value stream mapping—a method for identifying the inherent waste and losses within an operation analyse and design the flow of materials and information required.

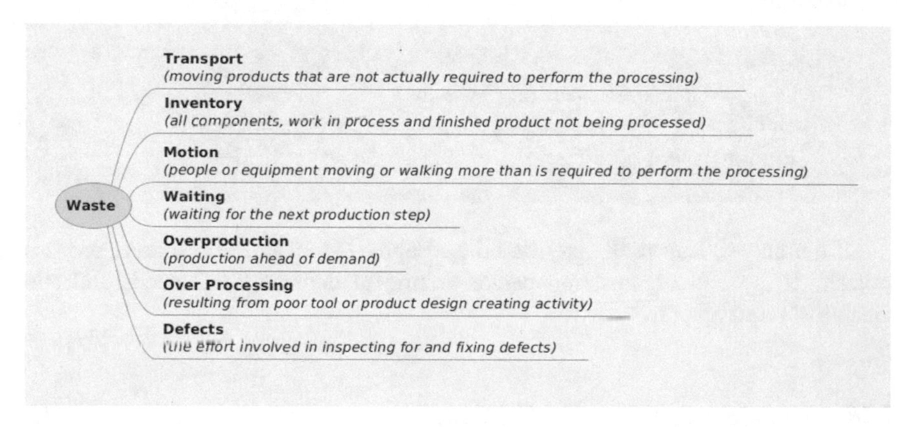

Fig. 6 Types of waste. *Source* https://en.wikipedia.org/wiki/Muda_(Japanese_term)#/media/File:Different_kinds_of_waste_in_lean_manufacturing.png

[95] https://www.greenbiz.com/news/2007/03/26/lean-manufacturing-environment-and-bottom-line.

[96] https://en.wikipedia.org/wiki/Lean_manufacturing.

6S—is well-known 5S plus the added step of safety.

Kanban—it is a visual signal to trigger an action as a layer on top of existing process towards gradual benefit.

Poka-yoke—error-proofing:

- Abnormality detection
- Stopping
- Correcting the immediate condition
- Root cause Investigation and installing countermeasure.

Total productive maintenance—elimination of time batching and mixed model processing.

Rank-order clustering—Grouping machines logically so that material handling can be minimised.

Single point scheduling[97]—This gives a real-time overview of customer's demand and to plan in advance, so that one can fulfil the requirements reliably and efficiently, while keeping finished stock inventories as low as possible.

KAIZEN[98]—All the stakeholders are given insights into the company's intentions, and they can warn the management if the stated sustainable targets have not been reached. As it is a continuous improvement, it is a good methodology to realise sustainability progressively.

The following steps ought to be implemented to create an ideal lean manufacturing:

- Design a simple manufacturing system
- Recognise that there is always room for improvement
- Continuously improve the lean manufacturing system design

All the above, lean tools may be utilised aptly based on the resource, process, product, utility and organisation nature linking to defence T&C needs and sustainability requirements.

5.10 Developments in T&C Front

The developments taking place in T&C front are most likely to be helpful for sustainability as discussed in the following paragraphs:

[97]http://www.industryweek.com/companies-amp-executives/lean-enterprise-goes-large.

[98]http://www.csrwire.com/csrlive/commentary_detail/2853-Kaizen-and-the-Art-of-Sustainability.

As communicated by Christer Wretfors Bengt Svennerstedt,[99] **Bio-fibres**, CO_2 neutral (when burned), from renewable materials tested for military purpose are most likely to become an important sustainable source in the conceptual age. With the emergence of nanotechnologies, textiles are benefitting a lot and the military textiles by the way of longer reliable performance will sure to become more sustainable.

Paul Anastas[100] has postulated twelve principles of **green chemistry**, which can be supportive to sustainable defence textile manufacturing. While environmental chemistry focuses on the effects of polluting chemicals on nature, green chemistry focuses on technological approaches to prevent pollution and reduce consumption of non-renewable resources.

As pointed out by Ingun Grimstad Klepp.[101] "**Better Mill Initiative**" is to develop tailor-made support to textile dyeing and finishing mills sustainability. "Zero-waste concepts", a thought, may conceivably contribute to the exertions taken towards sustainability. The effluent treatment processes such as biological treatment, coagulation-flocculation treatments (CFT), adsorption on activated carbon, ozone treatment, electrochemical technique, reverse osmosis, nano-filtration, and ultra-filtration used alone or combining (sometime as no process alone is efficient enough to treat the textile effluent and hence combination of one or two processes may be more effective) are expected to facilitate this inventiveness.

As proposed by "Natural Step"[102] **Sustainability Life Cycle Assessment (SLCA)** approach facilitates to define, assess and communicate product sustainability and provides a strategic outline on the social and ecological sustainability at the product level.

Life cycle mapping, assessment matrices, carbon foot printing, life cycle costing, energy mass balance, life cycle assessment, environmental life cycle assessment and social life cycle assessment as put forth by **PACIA**[103] as "**life cycle tool**" are apposite (for both new and experienced professionals) for sustainable approach.

As promoted by **SMART** criteria, "Sustainable Textile Standard"[104] is intended to allow inclusive participation and encourage the progressive drive of the textile industry towards sustainability. This standard identifies six levels of sustainable attribute performance and four levels of achievement by which textile materials and products can be measured with respect to specific attributes that link progress towards sustainability. "Sustainable Textile Supply Chain Achievement Matrix" covers all textile product stages: raw materials, transportation, manufacturing, use and final disposition. By using this standard, defence T&C supply chain

[99]http://allan.jbt.slu.se/publikationer/rapport/Rapport-142.pdf.

[100]https://en.wikipedia.org/wiki/Green_chemistry.

[101]http://norden.diva-portal.org/smash/get/diva2:840812/FULLTEXT01.pdf.

[102]http://www.thenaturalstep.org/our-work/slca/.

[103]http://supplychainsustainability.org.au/life_cycle_thinking/life_cycle_tools.

[104]https://www.google.co.in/?gfe_rd=cr&ei=ZjAbV6rtKcyo8wfJt7zQBA&gws_rd=ssl#q=SMART+criteria%2C+%E2%80%9CSustainable+Textile+Standard%E2%80%9D+.

stakeholders will be able to expressively and continuously improve their sustainable performance.

The benefits are:

- Design innovation by attentive contemplation of materials and resources
- Value added materials and products designed for safe reclamation and reuse
- Ecological restoration
- Long-term-reinforced customer relationships by offering forward-looking solutions to environmental problems
- A shared environmental programme serving local communities
- Reduced liability and need for regulation
- Compatibility with other market trends
- Government procurement and growing consumer demand for sustainable products.

As detailed by Jenna,[105] the reuse of polyester garments uses merely 1.8 % of the energy required for manufacture of these goods from virgin materials and the reuse of cotton clothing uses barely 2.6 % of the energy required to manufacture those from virgin materials.
Anne C. Wooldridge et al.,[106] says that taking into account extraction of resources, manufacture of materials, electricity generation, clothing collection, processing and distribution and final disposal of wastes, for every kilogram of virgin cotton displaced by second-hand clothing, approximately 65 kWh is saved, and for every kilogram of polyester, around 90 kWh is saved, hence to the advantage of economic and environmental sustainability.
As Ram Nidumolu et al.,[107] put forth that sustainability is mother layer of organisational and technological innovations that benefits both bottom line and top line by the way of reducing the inputs they use and generating additional revenues from sustainable products enabling companies to create new businesses.
Smart companies treat sustainability as innovation's new avant-garde. Derived from this, "Innovating sustainability"[108] means making intentional changes to

[105]Watson http://www.treehugger.com/style/974-net-energy-savings-from-reusing-cotton-clothing-lca-the-salvation-army.html.

[106]http://www.ecpar.org/sites/ecpar.org/files/documents/Wooldridge_LifeCycleAssessment.pdf.

[107]Nidumolu, R., Prahalad, C.K., &Rangaswami, M.R. 2009. Why sustainability is now a key driver of innovation. Harvard Business Review, 87(9): 57–64.

[108]http://nbs.net/wp-content/uploads/NBS-Executive-Report-Innovation.pdf.

organisational products or processes that produce environmental and/or social benefits as well as economic value and it has coined three paradigms:

(i) "Eco-Efficiency"—"Doing the same things better"
(ii) "New Market Opportunities"—"Doing good by doing new things"
(iii) "Societal Change" "Doing good by doing new things with others"

These three approaches are constituted as detailed below:

Eco-Efficiency—"Doing the same things better":

- Organisation level
- Product level
- Service level

New Market Opportunities—"Doing good by doing new things":

- Disruptive new products that change consumption habits
- Disruptive new products that benefit people
- Replacing products with services
- Replacing physical services with electronic services
- Services with social benefits

Societal Change—"Doing Good by Doing New Things with Others":

- Industrial symbiosis with the concept of "circular economy"

Sustainable design[109]—although the practical application varies among disciplines, some common philosophies such as low-impact materials, energy efficiency, emotionally durable, reusable, recyclable, material from renewable sources are expected to be futuristic thinking.

As cited by Thomas Bieker,[110] "Sustainability Balanced Score card (SBSC)" translates sustainability visions and strategies into action. It displays how intangible assets may contribute to the sustainability of companies. The SBSC initiates the integration of sustainable aspects and objectives into the core management.

[109]https://en.wikipedia.org/wiki/Sustainable_design.

[110]http://citeseerx.ist.psu.edu/viewdoc/download?rep=rep1&type=pdf&doi=10.1.1.200.9541.

The emerging three-dimensional printing, which uses ultraviolet beams to fuse layers of powdered, recyclable thermoplastic into shape, leaves behind virtually no waste. Its localised production and one-size-fits-all approach necessitates less labour and compresses fabrication.[111] A new bio-based nylon fibre with in-built moisture absorption and flame retardancy properties derived from renewable sources is being tested by the Chinese military to replace existing polyester-based fabrics likely to scale up the sustainability production.[112]
In a waterborne technology,[113] the PU is delivered and processed without solvents and provides a safer working environment; less pollution; and high efficiency, as textiles can be processed with 95 % less water and 50 % less energy compared with conventional technologies. A base layer of the coating is created from this.
Wal-Mart's[114] "three environmental goals" as signified below lead to immediate and far-reaching sustainability:

- To be supplied 100 per cent by renewable energy.
- To create zero waste.
- To sell products that sustain our resources and environment.

DyeCoo[115] sustainable route of waterless dyeing and supercritical CO_2 ($scCO_2$) textile processing technology is in offing to the benefit of textile processors.
"Closed loop" or "circular textiles",[116] in which new clothes are made from existing clothing and textiles, being practiced may have positive effect on sustainability.
The synthetic spider silk,[117] a forthcoming resource, can become a biodegradable soldier's protection for the conceptual era.

6 Best Practices

The best practices as elucidated beneath are ways forward to realise best sustainable defence T&C business practices.

---

[111]http://www.ecouterre.com/are-3d-printed-fabrics-the-future-of-sustainable-textiles/.

[112]Dornbirn, Chinese military backs bio−based nylon developments; Ecotextile News September 2015.

[113]http://www.technical-textiles.net/terms/coating-and-laminating-3.

[114]https://www.greenbiz.com/article/walmart-sustainability-10-birth-notion.

[115]http://www.huntsman.com/corporate/a/Innovation/DyeCoo%20delivers%20sustainable%20textiles.

[116]http://www.theguardian.com/sustainable-business/2014/sep/24/closed-loop–textile-recycling-technology-innovation.

[117]Moon parka, TechTex India, Jan−Mar 2016, vol. 10, issue 1.

NRDC[118] suggests ten best practices, viz., four water-saving best practices, five energy (Fuel)-saving best practices and one electricity-saving best practice.
DyeCoo's CO_2 technology,[119] 100 % water-free process and chemical-free solution with a lean and clean outlook being practiced by "Major Brands" is a major step forward in sustainable textile processing.
DryDye,[120] a breakthrough sustainable technology that altogether eliminates the need for water in the dyeing process uses 50 % less energy and 50 % fewer chemicals than traditional dyeing methods.
The ten pragmatic best practices for T&C industries sustainability practices according to Linda Greer et al.[121] are:

- Leak detection, preventive maintenance and improved cleaning
- Reuse cooling water
- Reuse condensate
- Reuse process water
- Recover heat from hot rinse water
- Pre-screen coal
- Maintain steam traps
- Insulate pipes, valves and flanges
- Recover heat from smokestacks and
- Optimise compressed air system

Defence authorities can take-up these measures for military T&C in support of sustainability;

As referred by Thomas Bieker,[122] "Sustainability Balanced Score card" concept offers an opportunity to translate sustainability visions and strategies into action. It demonstrates how intangible assets may contribute to the sustainability of companies. The SBSC provides high potential for the integration of environmental and social aspects and objectives into the core management.
Kelli and Sean Donovan suggest[123] that if entire industry teams together—that is from growers, manufacturers, designers, retailers, consumers, etc., T&C can make a transformation towards ethical and sustainable products.
Technology development in near future may result to freshen clothes without washing, efficient sorting of used clothing, new fibre recycling technology and new

[118]https://www.nrdc.org/sites/default/files/rsifullguide.pdf.
[119]http://www.dyecoo.com/dyecoo/.
[120]http://www.dyecoo.com/adidas-applies-drydye-technology-to-its-prime-t-range/.
[121]https://www.nrdc.org/sites/default/files/rsifullguide.pdf.
[122]http://citeseerx.ist.psu.edu/viewdoc/download?rep=rep1&type=pdf&doi=10.1.1.200.9541.
[123]http://www.ifm.eng.cam.ac.uk/uploads/Resources/Other_Reports/UK_textiles.pdf.

low-temperature detergents as a means to sustainability. Repurpose military surplus fabric into stylish purses and bags[124] is an example.
Wear2[125] process facilitates microwave textile disassembly, making it easy to debrand corporate clothing, remove labels from stock and prepare textiles such as clothing, car seats and mattresses for recycling. (*Until now, the lack of effective disassembly technologies and absence of design protocols for handling clothing at end-of-life have acted as a barrier in this regard.*)
Spider silks, biodegradable, stronger and tougher than steel, could be used for ballistic protection.[126]

7 Conclusion

The sustainable engineering is "the integration of social, environmental and economic considerations into product, process and energy system design methods". Industrialisation needs to be sustainable to ensure quality of life. The key sustainability challenges in fibre production[127] and in manufacturing practices to be addressed with a full-hearted outlook. Unsustainability elements had been set in motion as of no consequence inevitability in the early industrialisation process.

Defence T&C characterised with certain specific features met by a range of fibres, chemicals, and technical and management processes, which, for sustainability inevitability to be suitably look upon, encompassing the pillars of sustainability and duly accounting whole life cycle. Sustainable procurement process is capable of strengthening the sustainability phenomena. Tools and techniques as elucidated for twenty-first-century requirements are capable of substantiating sustainability efforts.

Some of the efforts as detailed below are positive approaches towards sustainable defence T&C[128]:

- Sourcing suppliers based on social and environmental performances;
- Substituting hazardous substances with safer substances;
- Inclusive information exchange with stake holders;
- Promoting more of sustainable fibres such as organic cotton, recycled fibres;

[124]http://daniellelvermeer.com/blog/upcycled-fashion-companies.

[125]http://www.ctechinnovation.com/funded-projects/wear2-microwave-textile-disassembly/.

[126]http://www.trustedclothes.com/blog/2016/05/18/the-future-of-sustainable-textiles/.

[127]Charlotte Turner, Reducing Environmental Impact, The Future of Fashion Fabrics—October 2012.

[128]http://ec.europa.eu/environment/industry/retail/pdf/issue_paper_textiles.pdf.

- Usage of various modus operandi capable of strengthening sustainability;
- Demanding suppliers to implement international social and environment standards;
- Using best practices.

The educational programmes on "Sustainable Management" and yearly "Textile Sustainability Conference" are capable of providing knowledge and skills on "green path" to transform the way that organisations do business for defence T&C.

Acknowledgments The authors thank the management and Dr. D. Saravanan, Principal of Bannari Amman Institute of Technology, and Prof. V Thanabal, HoD of Textile Technology, for their encouragement and facilitation to write this article.

Enzymatic Washing of Denim: Greener Route for Modern Fashion

Mohammad Shahid, Yuyang Zhou, Ren-Cheng Tang and Guoqiang Chen

Abstract Application of biotechnology in textiles is a fast developing area in textile industry. The textile research community has showed tremendous interest in developing enzymatic-based technologies for various steps in textile processing. Denim washing with enzymes is one of the most widely accepted enzyme-based techniques in textile industry. This chapter will give an overview of current research and developments in enzymatic denim washing process. A critical discussion on recent trends and future directions will be presented in order to illustrate benefits and limitations of such applications in sustainable development of future denim washing industry.

Keywords Biotechnology · Denim · Washing · Enzymes · Textile processing · Comfort · Visual appearance

1 Introduction

The growing application of biotechnologies in textile processing in last few decades has instrumental development in textile industry. The application of enzymes in processing of textile materials is growing day by day due to associated environmental benefits and mild processing conditions. Today, enzyme-based processes are being used in almost every major step in textile wet processing, such as desizing, scouring and bleaching of cotton, denim washing and biopolishing, bast fiber retting/degumming, wool scouring and shrink-proofing, silk degumming, biodye

M. Shahid · Y. Zhou · R.-C. Tang · G. Chen (✉)
National Engineering Laboratory for Modern Silk, College of Textile and Clothing Engineering, Soochow University, 199 Renai Road, Suzhou 215123, China
e-mail: chenguojiang@suda.edu.cn

S.S. Muthu (ed.), *Textiles and Clothing Sustainability*, Textile Science and Clothing Technology, DOI 10.1007/978-981-10-2474-0_3

production, enzyme-assisted dyeing, and finishing (Shahid et al. 2016; Araujo et al. 2008). According to a recent BCC research report, the global industrial enzyme market is expected to reach around $7.1 billion by 2018[1].

Denim garments are a major trendsetter in fast-changing fashion industry around the globe. Denim, usually made of thick coarse yarns of cotton, is a stiff fabric preferably dyed with indigo. Heavily starch sizing and very tight woven structure makes denim fabrics extremely sturdy and long-lasting material, but rather stiff and uncomfortable if not given a proper finishing treatment (Bhat 2000). A variety of denim garment treatment methods are being practiced for achieving smooth and soft hand feel (Kan et al. 2011). Until about 1974, most of the denim garments were sold without treatment and washing (Andreaus et al. 2001). Later on, in order to improve comfort properties and visual appearance with the market and fashion in mind, various mechanical and chemical methods were developed for denim fabrics treatment (Lee et al. 2015). A variety of desired effects and colors can be obtained on the same raw denim fabric by application of different dry and wet treatment processes such as scraping, washing, damaging, sanding, brushing, overdyeing, coating, tinting, fraying, printing, and embroidery (Kalaoglu and Paul 2015). Incredible effect of denim washing in making the fabric softer, suppler, smoother, and producing umpteen looks and effects makes it one of the most important steps denim garment preparation. It is an indispensable part of producing special effects on denim fashion garments.

2 Conventional Washing of Denims and Their Challenges

Over the time, the process of denim washing has significantly evolved and several techniques for producing unique color effects and washed looks were introduced such as regular washing, bleach washing, stone washing, enzyme washing, acid washing, sand blasting, monkey wash, brushing/grinding, whiskering, ozone fading, laser treatment, and water jet fading (Kan 2015). Traditional method of denim washing with pumice stones was developed in the late 1970s and early 1980s. In the stonewashing process, freshly dyed fabrics are washed with pumice stones to achieve a desirable look. The resultant effect greatly depends on variations in compositions, hardness, size shape, and porosity of pumice stones. The nonhomogeneous removal of dye from the fiber surface reveals the white interior of the yarn giving unique faded look. A wide range of specific effects can be obtained by using stone washing process alone or in combination with other processes (Shahid et al. 2016). However, damage to equipment and garments due to the overload of pumice stones and clogging of the machine drainage passage by particulate materials are serious drawbacks of this method (Yu et al. 2013). In the

[1]http://www.bccresearch.com/market-research/biotechnology/enzymes-industrial-applications-bio030h.html.

fast-changing textile industry, stakeholders are always looking for technological improvement to minimize resource consumption and maximize output. In the last few decades, advancement in denim washing techniques, particularly enzyme washing, led to dramatic improvement and superiority over former traditional methods. This is aimed to provide comprehensive information on development of enzyme wash process for denim garments, its limitations, and alternative solutions.

3 Denim Biowashing: Biotechnology for Modern Fashion

In the mid-1980s, the introduction of enzymes in denim washing revolutionized the process by providing a suitable alternative to stonewashing (Shahid et al. 2016; Bhat 2000; Belghith et al. 2001). The process was later known as 'biostoning' or 'biowashing.' During the last few decades, continuous research and development has established denim biowashing as one of the most exciting enzymatic processes in modern textile industry (Yu et al. 2013; Belghith et al. 2001; Cavaco-Paulo et al. 1998). Today, denim washing with cellulases has become a widely used practice and a variety of denim washing enzyme preparation are available in the market which can be used alone or in combination with other enzymes and pumice stones in order to obtain a specific look (Araujo et al. 2008; Choudhury 2014). A small amount of enzyme can replace several kilograms of stones and provide similar abrasive effect. The most widely used enzymes in denim washing are cellulases, although potential application of other enzymes such as laccases and amylases has also been presented by some researchers (Shahid et al. 2016).

3.1 *Cellulases—Ideal Enzymes for Denim Biowashing*

Cellulases, the hydrolytic enzymes that catalyze the breakdown of cellulose to smaller oligosaccharides and finally glucose, are synthesized by a large diversity of fungi, bacteria, protozoans, plants, and animals. Basically, cellulase preparations are multicomponent enzyme systems which act synergistic action of at least three types of cellulase: (i) endo-(1,4)-β-D-glucanase (EC 3.2.1.4), (ii) exo-(1,4)-β-D-glucanase (EC 3.2.1.91), and (iii) β-glucosidases (EC 3.2.1.21) (Araujo et al. 2008; Kuhad et al. 2011; Zhang and Zhang 2013). Endoglucanases break down cellulose chains in a random manner, cellobiohydrolases (exocellulases) liberate glucose dimers from both ends of the cellulose chains, and beta-glucosidases produce glucose from oligomer chains (Adrio and Demain 2014). Different types of cellulase preparations differ in their activity range of temperature and pH. Most of the cellulases are active in the range of 30–60 °C. Cellulases can be classified into three groups based on their pH activity range: acidic, neutral, or alkaline (Araujo et al. 2008). In the early 1980s, cellulases were first introduced for animal feed processing followed by food applications. Cellulases have become one of the most widely used enzymes due to their diverse industrial applications. Today, a dominant share of the world's industrial enzyme market is covered by cellulases due to its wider application in food and feed, textile, pulp and paper, and detergent industries. Cellulases have now become the third largest group of enzymes used in the textile and laundry because of their ability to modify cellulosic fibers in a controlled and desired manner (Bhat 2000). Cellulases find extensive usage in modern textile industry for the finishing of cellulosic fabrics, especially biopolishing (the removal of protruding surface fibrils to reduce pilling propensity) and biostoning (to achieve the worn-out look in denim garments) (Zilz et al. 2013). Several associated benefits of enzymatic biopolishing and biostoning include improved fabrics quality, improved absorbance property of fibers, softening of garments, improved stability of cellulosic fabrics, removal of excess dye from fabrics, and restoration of color brightness and fashionable looks (Kuhad et al. 2011). Commercial successes of denim biostoning and biopolishing of cellulosic fabrics are the best examples of cellulase application in textile industry. In the biostoning process, cellulases cause detachment of partially projecting dyed fiber from the fabric surface, thereby loosening the dye which is easily removed by mechanical abrasion in during washing, and reveal the underlying material (Kuhad et al. 2011; Rowe 1999). In traditional stonewashing process, this nonhomogeneous removal of dye from the fabric to obtain fashionable contrast of various shades is achieved with high levels of mechanical friction with pumice stones. The application of cellulases denim washing can reduce or even eliminate the use of stones, resulting in less damage to the garment and machine, and less pumice dust in the laundry environment. Productivity can also be increased because laundry machines contain fewer stones or none at all, and more garments (Araujo et al. 2008). Machine capacity can be improved by 30–50 % because of reduced processing time and better process control and lesser product variability (Rowe 1999). The technical, environmental, and economic benefits of enzymatic washing process led to gradual substitution of traditional stonewashing process by cellulase-based washing techniques. Since the first introduction of cellulases in the mid-1980s, denim biowashing process has evolved significantly. In modern denim industry, about 80 % of denim garments are washing using cellulase-based washing process (Tarhan and Sariisik 2009). A lot of research has been done to find cellulases with improved denim washing performance and different types of acidic, neutral, and alkaline cellulases and their mixtures are characterized and applied for denim washing (Shahid et al. 2016; Araujo et al. 2008; Belghith et al. 2001; Cavaco-Paulo et al. 1998; Cavaco-Paulo 1998; Cavaco-Paulo and Almeida 1994; Rau et al. 2008). The enzymatic washing involves several considerations with regard to the type of machinery used, garment properties, process parameters, and enzyme dosage to obtain desired look and effect (Puranen et al. 2014). Selections of appropriate pH and temperature conditions depending on the fabric type, required effect, and type of cellulase are very important factors to control the hydrolysis of cellulose by cellulases and minimize fabric strength loss (Kan et al. 2011; Šimić et al. 2015). Table 1 represents typical process parameters for an industrial biostone washing process.

Table 1 Typical process parameters for industrial biostoning. Reprinted from Puranen et al. 2014, Copyright (2014), with permission from Elsevier

Parameter	Value
pH	4.5–8.0
Temperature	30–55 °C
Incubation time	15–60 min
Liquor ratio	8:1–20:1
Enzyme dosage	0.5–3 % of weight of fabric
Auxiliary chemicals	Antibackstaining surfactants

Washing performance of different types of cellulose preparations differs significantly depending on their composition and washing conditions (Gusakov et al. 1998, 2000, 2001; Sinitsyn et al. 2001). As most of the commercial cellulose preparations are multicomponent systems, it is quite difficult to identify the reasons for variability of washing performance. *Trichoderma reesei* is one of the most widely used industrial cellulase producers because of ease, low cultivation cost, and safe production of large quantities of cellulases (Puranen et al. 2014). At *least* two different cellobiohydrolases (CBHs) and nine different endoglucanases (EGs) are produced by *T. reesei* (Miettinen-Oinonen et al. 2005). The cellulolytic efficiency of cellulases depends on synergistic action of both EGs and CBHS (Heikinheimo et al. 2000). Basically, EGs and CBHs consist of three distinct parts: (i) a cellulose binding domain helps the enzyme to attach on the cellulose surface, (ii) a core that catalyzes the cleavage, and (iii) a linker connecting the two regions (Fig. 1) (Colomera and Kuilderd 2015). The washing performance of cellulases is always applied in processes where strong mechanical action on the fabric is provided. The level of mechanical agitation greatly affects the relative activities of EG and CBH in a total crude mixture and thus plays an important role in determining the washing performance of cellulase preparations (Cavaco-Paulo 1998).

Heikinheimo et al. (2000) investigated the denim washing effects of three purified monocomponent cellulases, EG I, EG II, and CBH I, and two different cellulase mixtures produced by genetically modified strains of *T. reesei* by analyzing the soluble reducing sugars, absorbance, and lightness values of the treatment solutions. They reported that the purified cellulase EG II is most effective at

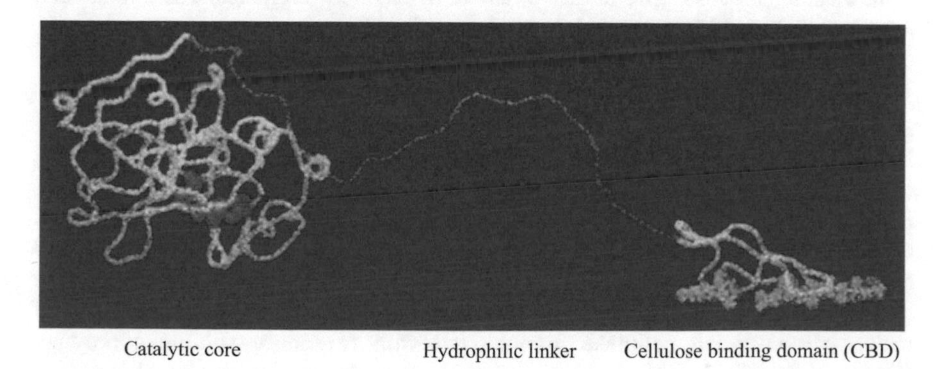

Fig. 1 Typical cellulase enzyme with binding domain. Reprinted from Colomera and Kuilderd 2015, Copyright (2015), with permission from Elsevier

removing color from denim, producing a good stone washing effect with the lowest hydrolysis level. Miettinen-Oinonen and Suominen (2002) achieved the same stonewashing effect with a considerably lower enzyme dosage using the EGII cellulase preparation derived from the EGII-overproducing strain than when using the parental strain. Cellulase preparations derived from the high EG II activity-producing strain showed better stonewashing effect than its parent strain at same enzyme dosage. Gusakov et al. (2000) compared the denim washing performance of six purified fungal cellulases, four endo-1,4-β-D-glucanases, and two cellobiohydrolases, using a model microassay. They found that a certain correlation exists between the washing performance of enzyme and quantity of aromatic residues exposed to solvent on the surface of protein globule or overall percentage of the surface hydrophobic residues (Gusakov et al. 2000). In addition to *T. reesei* cellulases, many other microbial sources have also been tested as source of cellulases for denim biowashing. Sinitsyn et al. used model microassays for testing the denim washing performance of cellulases from *T. reesei* and *Chrysosporium lucknowense*. *C. lucknowense* preparation demonstrated a higher potential in the denim biostoning process. Out of four purified *C. lucknowense* cellulases (two endoglucanases and two cellobiohydrolases), they identified that the endoglucanase with a molecular weight of 25 kDa was responsible for high abrasion effects on denim fabrics (Sinitsyn et al. 2001). Miettinen-Oinonen et al. (2004) tested culture supernatants from strains of *Melanocarpus albomyces*, *Myceliophthora thermophila*, *Chaetomium thermophilum*, and *Sporotrichum thermophilum* for their ability to release dye in neutral pH conditions from indigo-dyed cotton-containing fabric in biostoning applications. The supernatants from *M. albomyces* worked well in biostoning, with low backstaining. Out of the three purified cellulases (two endoglucanases with apparent molecular masses of 20 and 50 kDa and a 50 kDa cellobiohydrolase), 20 kDa endoglucanase performed well in biostoning of denim fabric at neutral pH and the addition of the purified 50 kDa endoglucanase or the 50 kDa cellobiohydrolase to the 20 kDa endoglucanase decreased backstaining in biostoning. Belghith et al. (2001) compared the action of the *Penicillium occitanis* (Pol6) cellulases with commercial cellulase (Novo) and traditional stonewashing process. Both types of cellulases gave better results than the traditional stonewashing process. The abrasive action of the Pol6 enzymes was similar and even better than with commercial cellulases (Fig. 2).

3.2 Challenges and Alternative Solutions

Although enzymatic washing of denim fabrics is a well-established technique, excessive hydrolysis of the cotton cellulose and the subsequent severe weight and tensile strength losses of the cotton fabric is a major concern associated with biowashing process. In a comparative study of denim washing performance of native cellulase and cellulase immobilized with reversibly soluble copolymer (Eudragit S-100), Yu et al. (2013) showed that the immobilized cellulase can

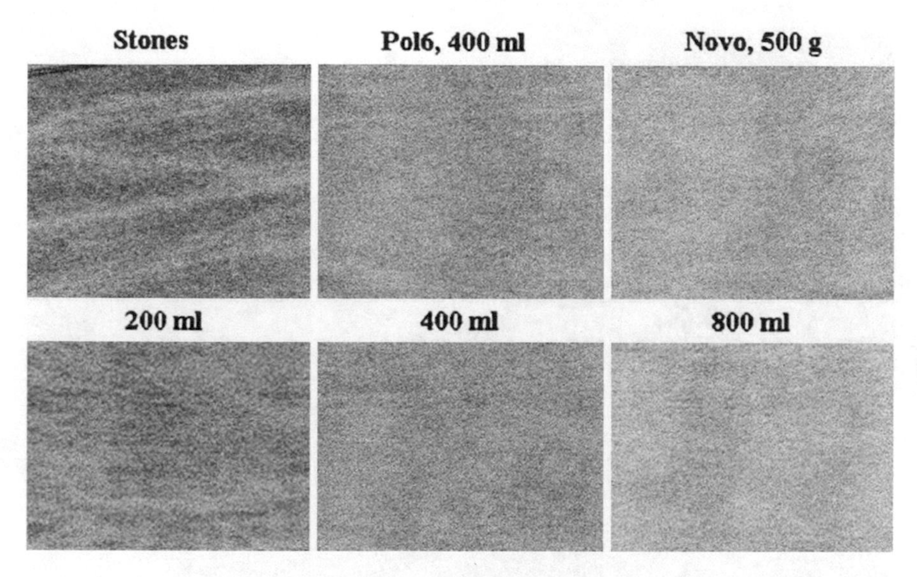

Fig. 2 Comparison between stonewashing and biostoning by commercial or Pol6 cellulases. Reprinted from Belghith et al. 2001, Copyright (2001), with permission from Elsevier

efficiently remove indigo dyestuffs from the surfaces of the denim fabrics without the problem of excessive damage to the fibers (Fig. 3). A cellulase concentration level of 6 % owf, the denim fabrics treated with the immobilized cellulase, showed decoloration and color effect close to the native cellulase. Recently, Lee et al. (2015) showed that a combination of liquid ammonia treatment with biowashing could provide the fade-out effect while maintaining softness, smoothness, good elastic recovery, and Koshi feel of the denim fabric.

The application of cellulases in denim washing is associated with another undesirable effect: backstaining, the redeposition of the indigo dyes on the white yarn of denim fabric (Sinitsyn et al. 2001). The high affinity of cellulase proteins to bind cotton cellulose is the main reason for backstaining problem (Cavaco-Paulo et al. 1998; Cavaco-Paulo 1998; Gusakov et al. 1998). Different types of cellulases differ significantly in their protein contents and thus show varied affinity for cellulosic substrates (Bayer et al. 1998). The backstaining increases with the increase in enzyme proteins adsorbed on fabric (Fig. 4) (Cavaco-Paulo et al. 1998). In order to prepare an ideal denim washing cellulase preparation, it must contain some sites capable of binding with indigo and lower affinity for cellulose (Shahid et al. 2016; Gusakov et al. 2001; Maryan and Montazer 2013). Gusakov et al. (1998) developed a microassay method for measuring indigo staining levels on white cotton fabric in the presence of enzymes which can be used for testing enzyme samples on a small laboratory scale and can predict the backstaining levels on a larger scale denim washing process. The correlations between the adsorption ability of enzymes and their backstaining index provide evidence that protein adsorption on cotton garment is a crucial parameter causing backstaining (Fig. 5).

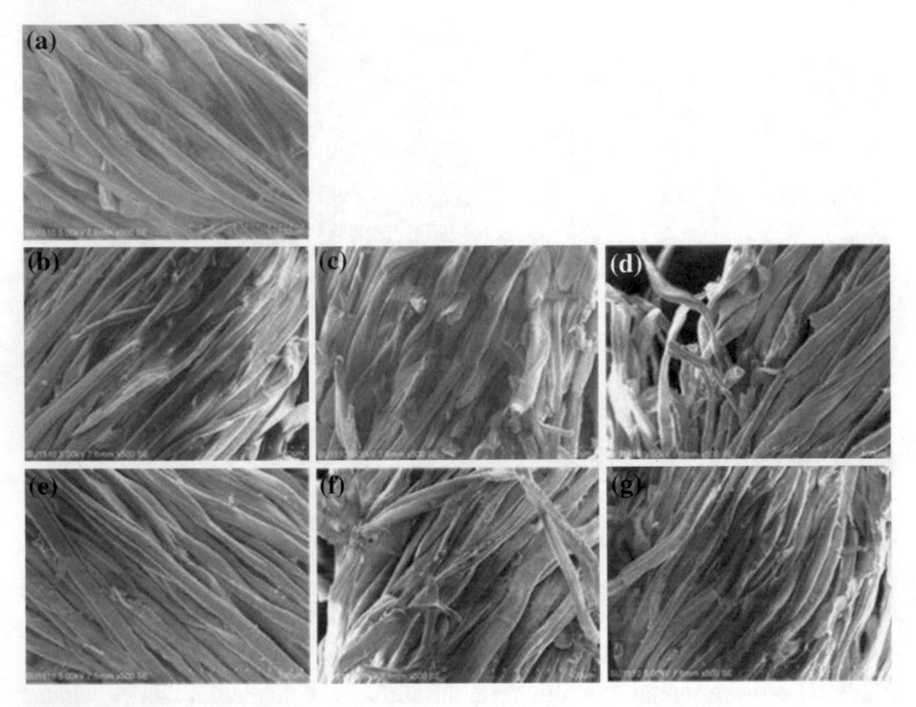

Fig. 3 SEM morphology of denim fabrics treated with **a** buffer (blank), **b** 2 % owf N, **c** 4 % owf N, **d** 6 % owf N, **e** 2 % owf I, **f** 4 % owf I, **g** 6 % owf I (*N* native cellulase, *I* immobilized cellulase). Reprinted from Yu et al. 2013, Copyright (2013), with permission from Elsevier

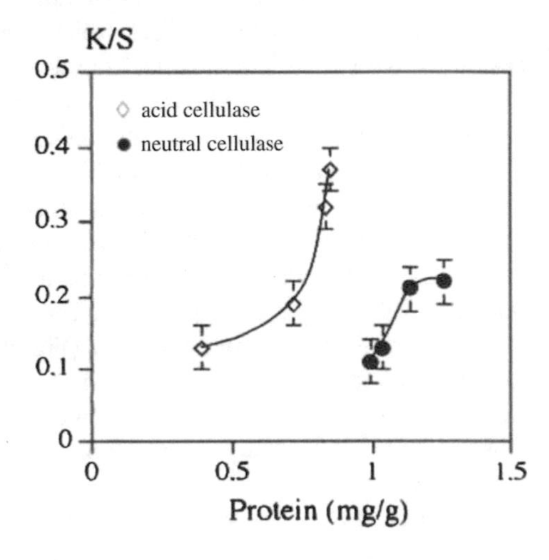

Fig. 4 Relationship between staining levels (*K*/*S*) and protein bound to the fabric. Reprinted from Cavaco-Paulo et al. 1998, Copyright (1998), with permission from SAGE Publications

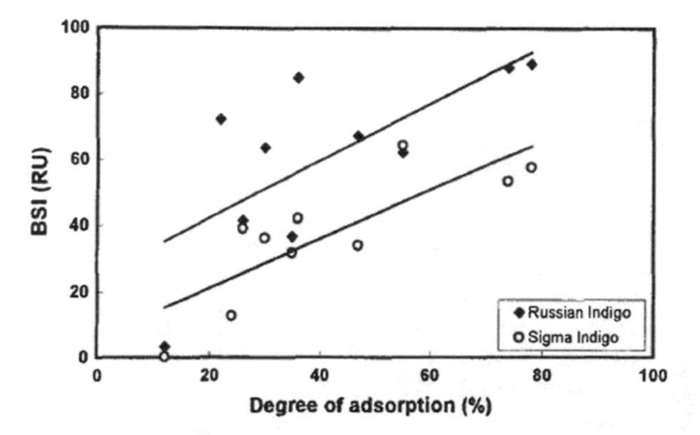

Fig. 5 Relationship between cellulase adsorption ability and the BSI at pH 5.0. Reprinted from Gusakov et al. 1998, Copyright (1999), with permission from Springer

Several measures have been proposed to minimize backstaining levels. The use of EGs devoid of their carbohydrate-binding domain (CBD) could be an alternative approach which could minimize the redeposition of dye molecules (Ferreira et al. 2014). Combined application of proteases along with cellulases in denim finishing was reported to achieve an improvement washing performance (Clarkson et al. 1994). Pretreatment of cellulosic fabrics with protease decreases the adsorbability of cellulase on cellulose resulting in lesser redeposition of indigo dye (Peng et al. 2005). Mamma et al. (2004) subjected a commercial cellulase preparation to limited proteolysis using papain in order to prevent redeposition of removed indigo dye during washing of cotton fabrics with cellulases. The limited proteolysis of cellulases reduced the endoglucanase activity (Fig. 6a) and resulted in preparation that exhibits lower efficiency of adsorption of cellulase to Avicel cellulose, and thus cause less backstaining (Fig. 6b).

Andreaus et al. (2001) proposed post-washing with different detergents, surfactants, and dispersing agents to remove indigo from stained cotton fabrics of cellulase-treated denim garments and reduce the undesired backstaining effect. Zilz et al. (2013) suggested simultaneous application of these auxiliaries together with cellulases, rather than the two-step procedure, to reduce process time and water consumption. An accurate selection of the type of cellulase or vigorous post-washing of the garments and the simultaneous application of auxiliaries such as surfactants and dispersing polymers during cellulase treatments on indigo- and pigment-dyed cotton fabrics can have additional benefits: (i) efficiently reduced backstaining of the dyes, and (ii) increased cellulase activity on the cellulosic substrate. Nature of the cellulase preparation used is also an important factor in determining backstaining level. By far, *Trichoderma* cellulases are most successful and widely used commercial enzymes for denim washing process (Puranen et al. 2014). However, there are several reports on new cellulase sources with better washing performance and lower backstaining. *Humicola insolens* cellulases (neutral) have been reported to have less backstaining compared to *T. reesei* cellulases (acid) (Cavaco-Paulo et al. 1998). Campos et al. (2000) studied the affinity of cellulases from different fungal origins for insoluble indigo dye and found that the acid cellulases from *T. reesei* have a higher affinity for indigo dye than neutral

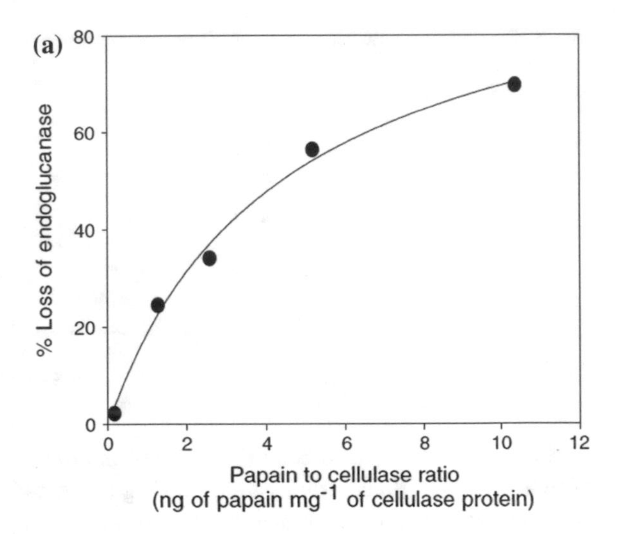

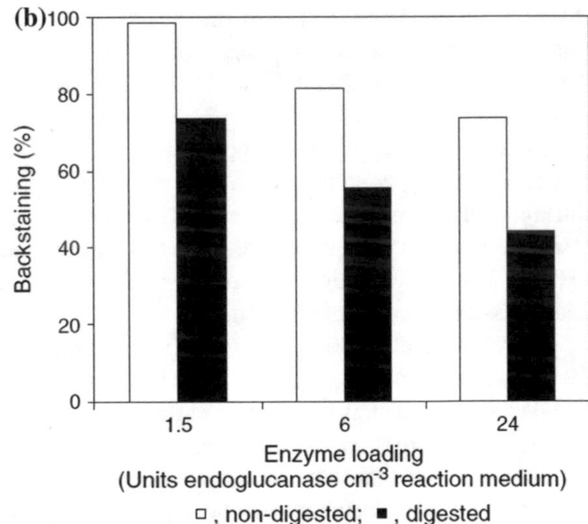

Fig. 6 **a** Effect of papain to cellulase ratio (ng of papain mg^{-1} of cellulase protein) on loss of endoglucanase activity. **b** Backstaining obtained with indigo for the non-digested Ecostone L350 and for the digested form at different endoglucanase concentrations. Reprinted from Mamma et al. 2004, Copyright (2004), with permission from John Wiley and Sons

cellulases of *H. insolens.* The particle size of indigo dye agglomerates is influenced by cellulase origin and concentration. The nonpolar residues present in higher percentages in the neutral cellulases of *H. insolens* seem to play an important role in the agglomeration of indigo dye particles and probably in the reduction of backstaining. Rau et al. (2008) reported that the cellulases from *Penicillium echinulatum* effectively remove more color from denim fabrics and produce less indigo dye redeposition (backstaining) than commercial acid or neutral cellulases.

Since cellulases are generally used for denim washing in soluble form, backstaining may also occur due to the cellulase–cellulose and indigo–cellulase affinities. The use of cellulase in immobilized form instead of soluble form could be a suitable alternative to solve this problem. The immobilization of cellulase on

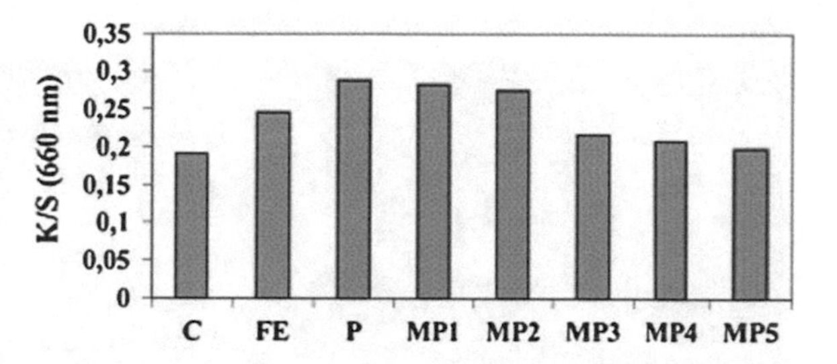

Fig. 7 *K/S* values of pocket material with treated control (*C*), pumice (*P*), free enzyme (*FE*), and immobilized acid cellulase (*MP*). Reprinted from Pazarlıoğlu et al. 2005b, Copyright (2005), with permission from Elsevier

pumice (Pazarlıoğlu et al. 2005b) and nanoclay (Maryan et al. 2015) for denim washing process is reported to have lower indigo staining on the denim garments. In addition, the immobilized cellulase can also reduce costs and raise the activity of the enzyme. The adsorption of the indigo dyes from washing effluent can significantly reduce the effluent color and backstaining. Pazarlıoğlu et al. (2005b) used $ZrOCl_2$-activated pumice as a carrier for the cellulase immobilization and compared the denim washing performance of immobilized acid cellulase with free enzymes and traditional denim washing procedure. Immobilized acid cellulases can efficiently abrade indigo-dyed denim fabrics (Fig. 7).

The application of laccases along with cellulases in washing bath is also reported to minimize the staining on white pocket and the back of denim with improved lightness and less fiber damage. In addition, laccases help to discolor the effluent, and enzyme residuals of both cellulases and laccases can be used for repeated processing with considerable reduction in consumption of the laccases, cellulases, water, time, and energy. At the optimum concentration of laccase/cellulase mixture, washing bath produces the same biowashing effect on the garment after three repeated biowashing (Montazer and Maryan 2008, 2010). The SEM pictures showed that fibers of treated sample with neutral cellulases damaged significantly, but fibers of treated sample with laccases and cellulases are little affected. In other words, laccases along with cellulases reduce the activity of neutral cellulases, and the fiber damages are limited to the outer fibers and inner fibers remain unaffected (Fig. 8).

The stone washing effect of indigo-dyed denim fabrics could be further improved by the laccase-mediated bleaching in order to obtain lighter shade (Pazarlıoğlu et al. 2005a). Montazer and Maryan (2010) investigated the effects of different enzymatic treatments including acid cellulases, neutral cellulases, and combination of laccases with cellulases and showed that the combination of laccases with cellulases help to improve the lightness and decrease staining on both back of garment and on white pocket without any significant variation of weight for different samples (Table 2). Laccases produced by white-rot fungi have been reported to bleach-dyed denim fabrics to lighter shades without causing any significant alteration to fabric weight and strength (Asgher et al. 2008). Due to the discoloration effect of laccase on dyes, it can be used along with cellulase in the biostoning of denim garment. It is able to degrade indigo both in solution and on denim, leading to various bleaching effects on the fabric (Maryan et al. 2013).

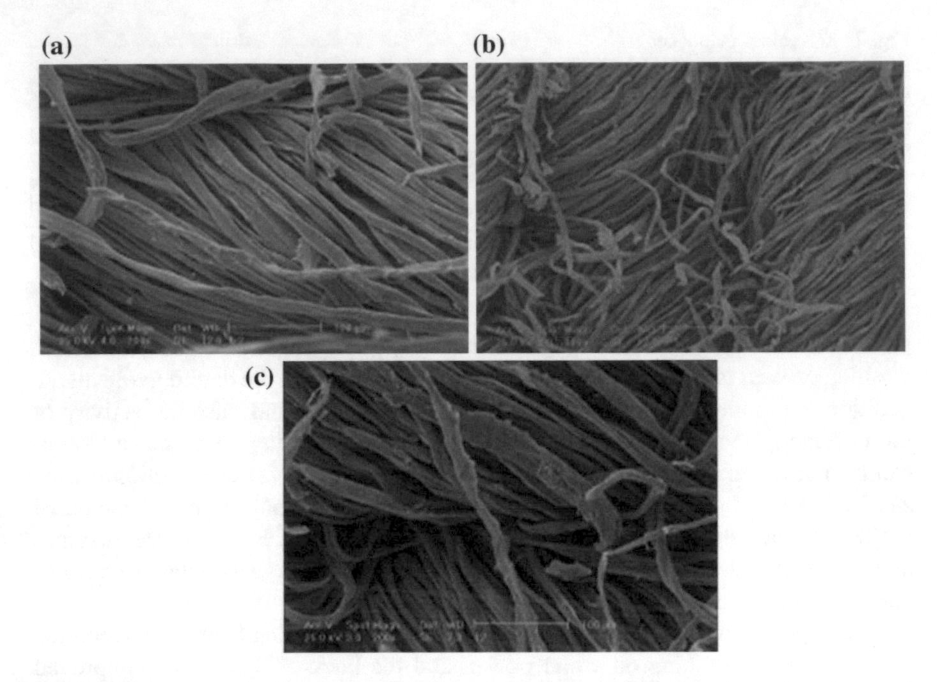

Fig. 8 SEM pictures of different treated samples, **a** desized sample, **b** treated sample with neutral cellulases, and **c** treated sample with mixture of cellulases and laccases. Reprinted from Montazer and Maryan 2008, Copyright (2008), with permission from John Wiley and Sons

Table 2 Percentage of weight loss for selected samples before and after abrasion. Reprinted from Montazer and Maryan 2010, Copyright (2009), with permission from Springer

Samples	Weight before abrasion (g)	Weight after abrasion (g)	Weight loss (%)
Desized sample	0.4978	0.4881	1.94
Treated sample with neutral cellulases	0.4977	0.4968	0.18
Treated sample with acid cellulases	0.4951	0.4923	0.55
Treated sample with laccases	0.5305	0.5239	1.24
Treated sample with mixture of laccases and acid cellulases	0.5359	0.5257	1.82
Treated sample with mixture of laccases and neutral cellulases	0.4945	0.4888	1.15

There are several stages involved in denim processing such as desizing, washing, enzyme treatment with cellulase, washing, and softening which require large amounts of water and energy. Maryan and Montazer (2013) proposed combined biodesizing and biowashing of denim garment in one step by using amylase, cellulase, and laccase together in order to reduce the water and energy consumption and increasing process efficiency. They reported that the non-desized samples treated with amylase/cellulase and laccase show similar properties of the samples

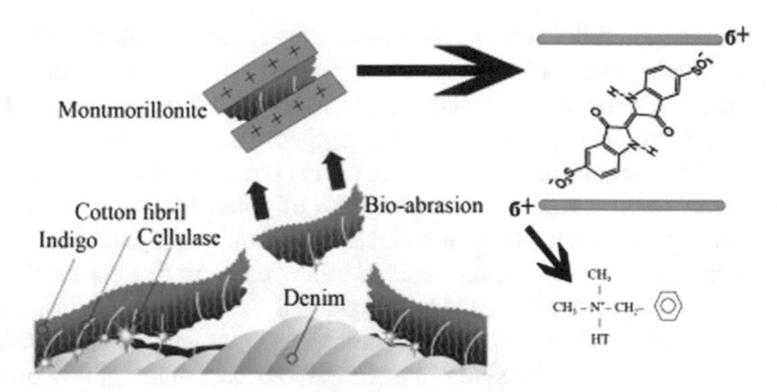

Fig. 9 Mechanism of enzymes and nanoclay action on denim fabric. Reprinted from Maryan et al. 2015, Copyright (2015), with permission from Elsevier

treated in two steps of biodesizing and biowashing. In addition, the fabrics treated with three enzymes showed lower backstaining with reasonable abrasion resistances up to 10,000 cycles. In another interesting study, Maryan et al. (2015) investigated the discoloration of indigo-dyed denim garment using a combination of bio- and nanotreatment (Fig. 9). Indigo-dyed denim garment was treated with enzymes in combination with modified montmorillonite-based nanoclay and mineral clay in the conventional washing machine. The application of enzymes along with montmorillonite-based nanoclay resulted in best aged-look effect without any significant mechanical damage to the fiber surface. The clay particle effectively adsorbed indigo dyes in washing liquor and consequently reduces backstaining.

4 Sustainability Aspects

Amid consumers' environmental attitudes toward eco-friendly sustainable fashion and growing industrial realization of the requirement of sustainable processing, denim industry are keenly trying to find new ways and processes which can help to minimize the environmental footprint of denim washing process. The evolution enzymatic method for denim washing purposes has led to the tremendous improvement with respect to performance at both economic and environmental considerations. Highly specific action; focused performance; reduced environmental impact cost reduction; reduced processing time, energy, and water savings; and improved quality and potential process integration of enzymatic washing are the basic reasons for acceptance of this technology by denim manufacturers. However, pressure is growing to minimize the water and energy consumption in order to provide more sustainable processing. More focused efforts are required to develop processes with reproducible effects, low temperature processing, reduced waste generation, and water consumption. In current practice, denim finishing with enzymes is usually performed in a batchwise process. Advances in biotechnology to create fast-acting and robust cellulases and development of special continuous-range equipment with greater mechanical action could make it possible to treat denim fabric in a continuous-process range in the future (Yoon 2005). The integration of newly available technologies laser, ozone, and e-flow together with new enzyme formulations and crosslinking agents could provide the possibility of producing denim fabrics with minimal quantities of water. Although much progress has been made in the development of cellulase compositions to achieve customized effects, continuous efforts are being made to improve the processing efficiency and better cellulase formulations. The development of genetic engineering tools has brought the opportunity to develop new cellulase enzymes with altered composition in order to provide higher abrasion contrast, reduced backstaining, and broadened operating pH and temperature ranges (Colomera and Kuilderd 2015). The development of new enzyme formulations and efficient strategies will further facilitate the denim industry to thrive with much higher throughput, better product quality, and a more sustainable processing strategy with minimized time, energy, and water consumption (Yoon 2005).

5 Conclusion and Future Perspectives

Application of biotechnology in textile industry is growing day by day. Enzyme-based processes are now integral part of modern textile industry and are utilized for a wide variety of applications. The application of enzymes in the denim washing industry is a well-established process for producing fashionable denim garments. The progress of the enzymatic technologies in denim washing has made it one of the most widely used washing processes. Cellulases are being used globally for denim washing. The factors affecting washing performance and problems associated with denim garment washing for practical usage and quality assurance are being actively monitored and researched both by academic and commercial stakeholders. Although the enzymatic denim washing process has evolved significantly in last two decades, there are still some challenges such as excessive fabric damage and backstaining. The development of innovative strategies to improve the processing conditions and finding more suitable enzyme formulations for enhanced process efficiency have been the focus of research in recent times. There have been a lot of different strategies proposed to counter the inherent drawbacks of cellulase washing process. Improvements in cellulase activities and imparting desired functionalities to enzymes through genetic engineering and finding new cellulase sources are the areas which should be investigated thoroughly. A more focused approach with engagement from key stakeholders in denim industry with the cooperation of academic researchers is required to find best-suited enzyme formulations, develop improved application methods, reduce the technical risks, and maximize economic and environment benefits.

Acknowledgments This study was funded by the Jiangsu Provincial Key Research and Development Program of China (BE2015066) and the Priority Academic Program Development (PAPD) of Jiangsu Higher Education Institutions.

References

Adrio, J. L., & Demain, A. L. (2014). Microbial enzymes: Tools for biotechnological processes. *Biomolecules, 4*(1), 117–139.

Andreaus, J., Campos, R., & Cavaco-Paulo, A. (2001). Reduction of indigo backstaining by post-washing. *Melliand International, 7*, 318–319.

Araujo, R., Casal, M., & Cavaco-Paulo, A. (2008). Application of enzymes for textile fibres processing. *Biocatalysis and Biotransformation, 26*(5), 332–349.

Asgher, M., Batool, S., Bhatti, H. N., Noreen, R., Rahman, S. U., & Asad, M. J. (2008). Laccase mediated decolorization of vat dyes by *Coriolus versicolor* IBL-04. *International Biodeterioration and Biodegradation, 62*(4), 465–470.

Bayer, E. A., Chanzy, H., Lamed, R., & Shoham, Y. (1998). Cellulose, cellulases and cellulosomes. *Current Opinion in Structural Biology, 8*(5), 548–557.

Belghith, H., Ellouz-Chaabouni, S., & Gargouri, A. (2001). Biostoning of denims by *Penicillium occitanis* (Pol6) cellulases. *Journal of Biotechnology, 89*(2–3), 257–262.

Bhat, M. K. (2000). Cellulases and related enzymes in biotechnology. *Biotechnology Advances, 18* (5), 355–383.

Campos, R., Cavaco-Paulo, A., Andreaus, J., & Gübitz, G. (2000). Indigo-cellulase interactions. *Textile Research Journal, 70*(6), 532–536.

Cavaco-Paulo, A. (1998). Mechanism of cellulase action in textile processes. *Carbohydrate Polymers, 37*(3), 273–277.

Cavaco-Paulo, A., & Almeida, L. (1994). Cellulase hydrolysis of cotton cellulose: The effects of mechanical action, enzyme concentration and dyed substrates. *Biocatalysis, 10*(1–4), 353–360.

Cavaco-Paulo, A., Morgado, J., Almeida, L., & Kilburn, D. (1998). Indigo backstaining during cellulase washing. *Textile Research Journal, 68*(6), 398–401.

Choudhury, A. K. R. (2014). Sustainable textile wet processing: Applications of enzymes. In S. S. Muthu (Ed.), *Roadmap to sustainable textiles and clothing* (pp. 203–238). Singapore: Springer.

Clarkson, K., Lad, P., Mullins, M., Simpson, C., Weiss, G., & Jacobs, L. (1994). *Enzymatic compositions and methods for producing a stone washed look on indigo-dyed denim fabric*. World patent PCT WO1994-29426.

Colomera, A., & Kuilderd, H. (2015). Biotechnological washing of denim jeans. In R. Paul (Ed.), *Denim manufacture, finishing and applications* (pp. 357–403). Amsterdam: Elsevier.

Ferreira, N., Margeot, A., Blanquet, S., & Berrin, J. (2014). Use of cellulases from *Trichoderma reesei* in the twenty-first century part I: Current industrial uses and future applications in the production of second ethanol generation. *Biotechnology and Biology of Trichoderma*, 245–261 (Elsevier, Oxford).

Gusakov, A. V., Sinitsyn, A. P., Berlin, A. G., Markov, A. V., & Ankudimova, N. V. (2000). Surface hydrophobic amino acid residues in cellulase molecules as a structural factor responsible for their high denim-washing performance. *Enzyme and Microbial Technology, 27* (9), 664–671.

Gusakov, A. V., Sinitsyn, A. P., Berlin, A. G., Popova, N. N., Markov, A. V., Okunev, O. N., et al. (1998). Interaction between indigo and adsorbed protein as a major factor causing backstaining during cellulase treatment of cotton fabrics. *Applied Biochemistry and Biotechnology, 75*(2–3), 279–293.

Gusakov, A. V., Sinitsyn, A. P., Markov, A. V., Sinitsyna, O. A., Ankudimova, N. V., & Berlin, A. G. (2001). Study of protein adsorption on indigo particles confirms the existence of enzyme–indigo interaction sites in cellulase molecules. *Journal of Biotechnology, 87*(1), 83–90.

Heikinheimo, L., Buchert, J., Miettinen-Oinonen, A., & Suominen, P. (2000). Treating denim fabrics with *Trichoderma reesei* cellulases. *Textile Research Journal, 70*(11), 969–973.

Kalaoglu, F., & Paul, R. (2015). *14—Finishing of jeans and quality control* (pp. 425–459). Denim: Woodhead Publishing.

Kan, C.W. (2015). 11—Washing techniques for denim jeans A2 (pp. 313–356). In R. Paul (Ed.) Denim: Woodhead Publishing.

Kan, C. W., Yuen, C. W. M., & Wong, W. Y. (2011). Optimizing color fading effect of cotton denim fabric by enzyme treatment. *Journal of Applied Polymer Science, 120*(6), 3596–3603.

Kuhad, R. C., Gupta, R., & Singh, A. (2011). Microbial cellulases and their industrial applications. *Enzyme Research, 2011*, 280696.

Lee, I. Y., Jeong, G. E., Kim, S. R., Bengelsdorff, C., & Kim, S. D. (2015). Effects of biowashing and liquid ammonia treatment on the physical characteristics and hand of denim fabric. *Coloration Technology, 131*(3), 192–199.

Mamma, D., Kalantzi, S. A., & Christakopoulos, P. (2004). Effect of adsorption characteristics of a modified cellulase on indigo backstaining. *Journal of Chemical Technology and Biotechnology, 79*(6), 639–644.

Maryan, A. S., & Montazer, M. (2013). A cleaner production of denim garment using one step treatment with amylase/cellulase/laccase. *Journal of Cleaner Production, 57*, 320–326.

Maryan, A. S., Montazer, M., & Damerchely, R. (2015). Discoloration of denim garment with color free effluent using montmorillonite based nano clay and enzymes: Nano bio-treatment on denim garment. *Journal of Cleaner Production, 91*, 208–215.

Maryan, A. S., Montazer, M., Harifi, T., & Rad, M. M. (2013). Aged-look vat dyed cotton with anti-bacterial/anti-fungal properties by treatment with nano clay and enzymes. *Carbohydrate Polymers, 95*(1), 338–347.

Miettinen-Oinonen, A., Londesborough, J., Joutsjoki, V., Lantto, R., & Vehmaanperä, J. (2004). Three cellulases from *Melanocarpus albomyces* for textile treatment at neutral pH. *Enzyme and Microbial Technology, 34*(3–4), 332–341.

Miettinen-Oinonen, A., Paloheimo, M., Lantto, R., & Suominen, P. (2005). Enhanced production of cellobiohydrolases in *Trichoderma reesei* and evaluation of the new preparations in biofinishing of cotton. *Journal of Biotechnology, 116*(3), 305–317.

Miettinen-Oinonen, A., & Suominen, P. (2002). Enhanced production of *Trichoderma reesei* endoglucanases and use of the new cellulase preparations in producing the stonewashed effect on denim fabric. *Applied and Environmental Microbiology, 68*(8), 3956–3964.

Montazer, M., & Maryan, A. S. (2008). Application of laccases with cellulases on denim for clean effluent and repeatable biowashing. *Journal of Applied Polymer Science, 110*(5), 3121–3129.

Montazer, M., & Maryan, A. S. (2010). Influences of different enzymatic treatment on denim garment. *Applied Biochemistry and Biotechnology, 160*(7), 2114–2128.

Pazarlıoğlu, N. K., Sariişik, M., & Telefoncu, A. (2005a). Laccase: Production by *Trametes versicolor* and application to denim washing. *Process Biochemistry, 40*(5), 1673–1678.

Pazarlıoğlu, N. K., Sariişik, M., & Telefoncu, A. (2005b). Treating denim fabrics with immobilized commercial cellulases. *Process Biochemistry, 40*(2), 767–771.

Peng, Z., Zhi-Hua, K., & Jian-Fu, T. (2005). Relationship between cellulase and Indigo dyes. *Dyeing & Finishing*, 11–14.

Puranen, T., Alapuranen, M., & Vehmaanperä, J. (2014). Chapter 26—Trichoderma enzymes for textile industries. In Biotechnology and biology of Trichoderma (pp. 351–362). Amsterdam: Elsevier.

Rau, M., Heidemann, C., Pascoalin, A. M., Filho, E. X. F., Camassola, M., Dillon, A. J. P., et al. (2008). Application of cellulases from *Acrophialophora nainiana* and *Penicillium echinulatum* in textile processing of cellulosic fibres. *Biocatalysis and Biotransformation, 26*(5), 383–390.

Rowe, H. D. (1999). Biotechnology in the textile/clothing industry—A review. *Journal of Consumer Studies & Home Economics, 23*(1), 53–61.

Shahid, M., Mohammad, F., Chen, G., Tang, R.-C., & Xing, T. (2016). Enzymatic processing of natural fibres: White biotechnology for sustainable development. *Green Chemistry, 18*, 2256–2281.

Šimić, K., Soljačić, I., & Pušić, T. (2015). Application of cellulases in the process of finishing Uporaba celulaz v procesu plemenitenja.

Sinitsyn, A. P., Gusakov, A. V., Grishutin, S. G., Sinitsyna, O. A., & Ankudimova, N. V. (2001). Application of microassays for investigation of cellulase abrasive activity and backstaining. *Journal of Biotechnology, 89*(2–3), 233–238.

Tarhan, M., & Sariisik, M. (2009). A comparison among performance characteristics of various denim fading processes. *Textile Research Journal, 79*(4), 301–309.

Yoon, M.-Y. (2005). Denim Finishing with Enzymes. *Dyer International, 2005*(11), 16–19.

Yu, Y., Yuan, J., Wang, Q., Fan, X., Ni, X., Wang, P., et al. (2013). Cellulase immobilization onto the reversibly soluble methacrylate copolymer for denim washing. *Carbohydrate Polymers, 95* (2), 675–680.

Zhang, X.-Z., & Zhang, Y.-H. P. (2013). *Cellulases: Characteristics, sources, production, and applications*. New Jersey: Wiley.

Zilz, L., Rau, M., Budag, N., Scharf, M., Cavaco-Paulo, A., & Andreaus, J. (2013). Nonionic surfactants and dispersants for biopolishing and stonewashing with *Hypocrea jecorina* cellulases. *Coloration Technology, 129*(1), 49–54.

Printed in the United States
By Bookmasters